PREFACE

The use of the differential geometry of a Riemannian space in the mathematical formulation of recent physical theories led to important developments in the geometry of such spaces. The concept of parallelism of vectors, as introduced by Levi-Civita, gave rise to a theory of the affine properties of a Riemannian space. Covariant differentiation, as developed by Christoffel and Ricci, is a fundamental process in this theory. Various writers, notably Eddington, Einstein and Weyl, in their efforts to formulate a combined theory of gravitation and electromagnetism, proposed a simultaneous generalization of this process and of the definition of parallelism. This generalization consisted in using general functions of the coördinates in the formulas of covariant differentiation in place of the Christoffel symbols formed with respect to the fundamental tensor of a Riemannian space. This has been the line of approach adopted also by Cartan, Schouten and others. When such a set of functions is assigned to a space it is said to be affinely connected.

From the affine point of view the geodesics of a Riemannian space are the straight lines, in the sense that the tangents to a geodesic are parallel with respect to the curve. In any affinely connected space there are straight lines, which we call the paths. A path is uniquely determined by a point and a direction or by two points within a sufficiently restricted region. Conversely, a system of curves possessing this property may be taken as the straight lines of a space and an affine connection deduced therefrom. This method of departure was adopted by Veblen and the writer in their papers dealing with the geometry of paths, the equations of the paths being a generalization of those of geodesics by the process described in the first paragraph.

In presenting the development of these ideas we begin with a definition of covariant differentiation which involves functions L^i_{jk} of the coördinates, the law connecting the corresponding functions in any two coördinate systems being fundamental. Upon this foundation a general tensor calculus is built and a theory of parallelism.

Much of the literature on this subject deals with the case where the connection is symmetric, that is $L^i_{jk} = L^i_{kj}$. When the paths are taken as fundamental, this is the type of connection which is derived. This restriction is not made in the first chapter, which deals accordingly with asymmetric connections.

Vectors parallel with respect to a curve for an asymmetric connection retain this property for certain changes of the connection. This is not true of symmetric connections. However, it is possible to change a symmetric connection without changing the equations of the paths of the manifold. Accordingly when the paths are taken as fundamental, the affine connection is not uniquely defined, and we have a group of affine connections with the same paths, a situation analogous to that in the projective geometry of straight lines. Accordingly there is a projective geometry of paths dealing with that theory which applies to all affine connections with the same paths. In the second chapter we develop the affine theory of symmetric connections and in the third chapter the projective theory.

For a sub-space of a Riemannian space there is in general an induced metric and consequently an induced law of parallelism. There is not a unique induced affine connection in a sub-space of an affinely connected space. If the latter is of order m and the sub-space of order n, each choice at points of the latter of $m-n$ independent directions in the enveloping space but not in the sub-space leads to an induced affine connection, and to a geometry of the subspace in many ways analogous to that for Riemannian geometry. Under certain conditions there are preferred choices of these directions, which are analogous to the normals to

NON-RIEMANNIAN GEOMETRY

Luther Pfahler Eisenhart

DOVER PUBLICATIONS, INC.
Mineola, New York

Bibliographical Note

This Dover edition, first published in 2005, is an unabridged republication of the work originally published by the American Mathematical Society in 1927.

Library of Congress Cataloging-in-Publication Data

Eisenhart, Luther Pfahler, b. 1876.
Non-Riemannian geometry / Luther Pfahler Eisenhart.
p. cm.
Originally published: New York : American Mathematical Society, 1927, in series: American Mathematical Society colloquium publications, v. 8.
Includes bibliographical references.
ISBN 0-486-44243-8 (pbk.)
1. Geometry, Differential. I. Title.

QA641.E55 2005
516.3'6—dc22

2004065109

Manufactured in the United States of America
Dover Publications, Inc., 31 East 2nd Street, Mineola, N.Y. 11501

the sub-space. The fourth chapter of the book deals with the geometry of sub-spaces.

A generalization of Riemannian spaces other than those presented in this book consists in assigning to the space a metric based upon an integral whose integrand is homogeneous of the first degree in the differentials. Developments of this theory have been made by Finsler, Berwald, Synge and J. H. Taylor. In this geometry the paths are the shortest lines, and in that sense are a generalization of geodesics. Affine properties of these spaces are obtained from a natural generalization of the definition of Levi-Civita for Riemannian spaces. Berwald has also obtained generalizations of the geometry of paths by taking for the paths the integral curves of a certain type of differential equations, and Douglas showed that these are the most general geometries of paths; he also developed their projective theory. References to the works of these authors are to be found in the Bibliography at the end of the book.

This book contains, with subsequent developments, the material presented in my lectures at the Ithaca Colloquium, in September 1925, under the title The New Differential Geometry. I have given the book a more definitive title.

In the preparation of the manuscript I have had the benefit of suggestions and criticisms by Dr. Harry Levy, Dr. J. M. Thomas and Mr. M. S. Knebelman, the latter of whom has also read the proof.

September, 1927.

LUTHER PFAHLER EISENHART.
Princeton University

CONTENTS

CHAPTER I

ASYMMETRIC CONNECTIONS

CHAPTER II

SYMMETRIC CONNECTIONS

CHAPTER III

PROJECTIVE GEOMETRY OF PATHS

CHAPTER IV

THE GEOMETRY OF SUB-SPACES

CHAPTER I

ASYMMETRIC CONNECTIONS

1. Transformation of coördinates. Any ordered set of n independent real variables x^i, where i takes the values $1, \cdots, n$, may be thought of as coördinates of points in an n-dimensional *space* V_n in the sense that each set of values of the x's defines a point of V_n. The terms *manifold* and *variety* are synonymous with space as here defined. If $\varphi^i(x^1, \cdots, x^n)$ for $i = 1, \cdots, n$ are real functions, whose jacobian is not identically zero, the equations

$$x'^i = \varphi^i(x^1, \cdots, x^n) \qquad (i = 1, \cdots, n) \tag{1.1}$$

define a transformation of coördinates in the space V_n.

If λ^i and λ'^i are functions of the x's and x''s such that

$$\lambda^i = \lambda'^\alpha \frac{\partial x^i}{\partial x'^\alpha} \tag{1.2}$$

in consequence of (1.1), λ^i and λ'^i are the components in the respective coördinate systems of a contravariant vector. In (1.2) we make use of the convention that when the same index appears as a subscript and superscript in a term this term stands for the sum of the terms obtained by giving the index each of its n values; this convention will be used throughout the book. From (1.2) we have by differentiation

$$\frac{\partial \lambda^i}{\partial x^j} = \frac{\partial \lambda'^\alpha}{\partial x'^\beta} \frac{\partial x^i}{\partial x'^\alpha} \frac{\partial x'^\beta}{\partial x^j} + \lambda'^\alpha \frac{\partial^2 x^i}{\partial x'^\alpha \partial x'^\beta} \frac{\partial x'^\beta}{\partial x^j}. \tag{1.3}$$

It is assumed that the reader is familiar with relations connecting the components of a tensor in two coördinate systems.*

* Cf. 1926, 1, pp. 1–12. References are to the Bibliography at the end of the text.

He will observe that, because of the presence of the last term in the right-hand member of (1.3), the derivatives of λ^i and λ'^i are not the components of a tensor.

Consider further a symmetric covariant tensor of the second order whose components in the two coördinate systems are g_{ij} and $g'_{\alpha\beta}$ such that the determinant

$$g = |g_{ij}| \tag{1.4}$$

is different from zero. From the equations

$$g'_{\alpha\beta} = g_{ij}\frac{\partial x^i}{\partial x'^\alpha}\frac{\partial x^j}{\partial x'^\beta}$$

we get by differentiation

$$\frac{\partial g'_{\alpha\beta}}{\partial x'^\gamma} = \frac{\partial g_{ij}}{\partial x^k}\frac{\partial x^i}{\partial x'^\alpha}\frac{\partial x^j}{\partial x'^\beta}\frac{\partial x^k}{\partial x'^\gamma} + g_{ij}\left(\frac{\partial x^i}{\partial x'^\alpha}\frac{\partial^2 x^j}{\partial x'^\beta\,\partial x'^\gamma} + \frac{\partial x^j}{\partial x'^\beta}\frac{\partial^2 x^i}{\partial x'^\alpha\,\partial x'^\gamma}\right).$$

A similar observation applies to these equations. However, there are $n^2(n+1)/2$ of these equations, and they can be solved for the $n^2(n+1)/2$ quantities $\dfrac{\partial^2 x^i}{\partial x'^\alpha\,\partial x'^\beta}$. We obtain

$$\frac{\partial^2 x^i}{\partial x'^\alpha\,\partial x'^\beta} + \begin{Bmatrix} i \\ jk \end{Bmatrix}\frac{\partial x^j}{\partial x'^\alpha}\frac{\partial x^k}{\partial x'^\beta} = \begin{Bmatrix} \gamma \\ \alpha\,\beta \end{Bmatrix}'\frac{\partial x^i}{\partial x'^\gamma},\text{*} \tag{1.5}$$

where $\begin{Bmatrix} i \\ jk \end{Bmatrix}$ are the Christoffel symbols of the second kind, that is,

$$\begin{Bmatrix} i \\ jk \end{Bmatrix} = g^{ih}[jk, h], \quad [jk, h] = \frac{1}{2}\left(\frac{\partial g_{jh}}{\partial x^k} + \frac{\partial g_{kh}}{\partial x^j} - \frac{\partial g_{jk}}{\partial x^h}\right), \tag{1.6}$$

where g^{ih} are defined by

$$g^{ih}g_{jh} = \delta^i_j, \qquad \delta^i_j = 1 \text{ or } 0 \text{ as } i = j \text{ or } i \neq j. \tag{1.7}$$

When we eliminate $\dfrac{\partial^2 x^i}{\partial x'^\alpha\,\partial x'^\beta}$ from (1.3) by means of (1.5), we obtain

* 1926, 1, p. 19.

$$(1.8) \qquad \lambda^i_{,j} = \lambda'^\alpha_{,\beta} \frac{\partial x^i}{\partial x'^\alpha} \frac{\partial x'^\beta}{\partial x^j},$$

where

$$(1.9) \quad \lambda^i_{,j} = \frac{\partial \lambda^i}{\partial x^j} + \lambda^h \left\{ {i \atop h\, j} \right\}, \qquad \lambda'^\alpha_{,\beta} = \frac{\partial \lambda'^\alpha}{\partial x'^\beta} + \lambda'^\gamma \left\{ {\alpha \atop \gamma\, \beta} \right\}'$$

From (1.8) we see that $\lambda^i_{,j}$ and $\lambda'^\alpha_{,\beta}$ are the components of a tensor in the two coördinate systems. Thus we have formed a tensor by suitable combinations of the first derivatives of the components of a vector and a tensor.

If g_{ij} is the fundamental tensor of a Riemannian space, then $\lambda^i_{,j}$ is the covariant derivative of λ^i. However, the theory of covariant differentiation in a Riemannian space has nothing to do with the fact that the tensor g_{ij} is used to define a metric. Consequently this theory can be applied to any space, if we make use of any tensor g_{ij} such that $g \neq 0$.*

2. Coefficients of connection. We have just seen that when a symmetric tensor g_{ij} is specified for a space we have an algorithm for obtaining tensors from other tensors by differentiation. But this process is a special case of a much more general one. In fact, the fundamental element in the former consisted in the elimination of $\dfrac{\partial^2 x^i}{\partial x'^\alpha \partial x'^\beta}$ from equations (1.3) by means of (1.5). From this it is evident that if L^i_{jk} and $L'^\gamma_{\alpha\beta}$ are functions of the x's and x''s satisfying the equations

$$(2.1) \qquad \frac{\partial^2 x^i}{\partial x'^\alpha \partial x'^\beta} + L^i_{jk} \frac{\partial x^j}{\partial x'^\alpha} \frac{\partial x^k}{\partial x'^\beta} = L'^\gamma_{\alpha\beta} \frac{\partial x^i}{\partial x'^\gamma},$$

the quantities $\lambda^i_{|j}$ and $\lambda'^\alpha{}_{|\beta}$, where

$$(2.2) \quad \lambda^i_{|j} = \frac{\partial \lambda^i}{\partial x^j} + \lambda^h L^i_{hj}, \qquad \lambda'^\alpha_{|\beta} = \frac{\partial \lambda'^\alpha}{\partial x'^\beta} + \lambda'^\gamma L'^\alpha_{\gamma\beta},$$

are in the relations

$$(2.3) \qquad \lambda^i_{|j} = \lambda'^\alpha_{|\beta} \frac{\partial x^i}{\partial x'^\alpha} \frac{\partial x'^\beta}{\partial x^j}$$

and consequently are components of a tensor.

* Cf. 1926, 1, pp. 26–30.

Conversely, if equations (2.3) are to hold for any vector, it follows from (1.3) that we must have

$$\frac{\partial^2 x^i}{\partial x'^\alpha \partial x'^\beta}\frac{\partial x'^\beta}{\partial x^j} + L^i_{kj}\frac{\partial x^k}{\partial x'^\alpha} = \frac{\partial x^i}{\partial x'^\gamma}\frac{\partial x'^\beta}{\partial x^j} L'^\gamma_{\alpha\beta}$$

which are equivalent to (2.1), since

$$(2.4) \qquad \frac{\partial x'^\beta}{\partial x^j}\frac{\partial x^j}{\partial x'^\alpha} = \delta^\beta_\alpha, \qquad \frac{\partial x^j}{\partial x'^\alpha}\frac{\partial x'^\alpha}{\partial x^k} = \delta^j_k.$$

If we take any set of functions L^i_{jk} of the x's, equations (2.1) determine the corresponding functions in any other coördinate system x'^i such that equations (2.2) define the components of the same tensor in the two coördinate systems. The particular form of the functions L^i_{jk} in (1.5) arose from a tensor g_{ij}, and there are other ways (cf. § 18) in which we get functions L^i_{jk} and $L'^\gamma_{\alpha\beta}$ in two coördinate systems satisfying (2.1). Whenever in any way such a set of functions is assigned to a manifold we say that the latter is *connected* and that the L's are the *coefficients of the connection.*

From (1.6) it is seen that the symbols $\left\{ {i \atop jk} \right\}$ are symmetric in j and k. We remark that from the form of (2.1) it follows that, if the L's are symmetric in the subscripts in one coördinate system, the corresponding coefficients in any coördinate system are symmetric. We do not make the restriction that they be symmetric, and for the present consider the more general case where the connection is *asymmetric.* Cartan* uses the terms *with torsion* and *without torsion* for the asymmetric and symmetric connections respectively.

When we express the conditions of integrability of equations (2.1), making use of (2.1) in the reduction, we obtain

$$(2.5) \qquad L^i_{jkl}\frac{\partial x^j}{\partial x'^\alpha}\frac{\partial x^k}{\partial x'^\beta}\frac{\partial x^l}{\partial x'^\gamma} = L'^\sigma_{\alpha\beta\gamma}\frac{\partial x^i}{\partial x'^\sigma},$$

* 1923, 5, pp. 325, 326.

where

$$L^i_{jkl} = \frac{\partial L^i_{jl}}{\partial x^k} - \frac{\partial L^i_{jk}}{\partial x^l} + L^h_{jl} L^i_{hk} - L^h_{jk} L^i_{hl}, \tag{2.6}$$

and similarly for $L'^{\sigma}_{\alpha\beta\gamma}$. From the form of (2.5) it follows that L^i_{jkl} and $L'^{\sigma}_{\alpha\beta\gamma}$ are the components of a tensor. Also if in (2.6) the functions L^i_{jk} are replaced by the Christoffel symbols $\left\{ {i \atop jk} \right\}$ formed with respect to g_{ij}, the tensor L^i_{jkl} becomes the Riemannian curvature tensor of a Riemannian space with the fundamental tensor g_{ij}.* Accordingly we call L^i_{jkl} the *curvature tensor* of the space.†

3. Covariant differentiation with respect to the L's. Since (2.2) are a generalization of (1.9), we call the tensor of components $\lambda^i{}_{|k}$ the *first covariant derivative of λ^i with respect to the given connection*, or briefly, *with respect to the L's*.

If λ_i are the components of a covariant vector-field, it is readily shown by means of (2.1) that the quantities $\lambda_{i|j}$, given by

$$\lambda_{i|j} = \frac{\partial \lambda_i}{\partial x^j} - \lambda_l L^l_{ij}, \tag{3.1}$$

are the components of a tensor of the second order. It is the *first covariant derivative* of the vector λ_i *with respect to the L's.*

In general it can be shown that, if $a^{r_1\cdots r_m}_{s_1\cdots s_p}$ are the components of a tensor, the quantities

$$\begin{aligned} a^{r_1\cdots r_m}_{s_1\cdots s_p|i} = \frac{\partial a^{r_1\cdots r_m}_{s_1\cdots s_p}}{\partial x^i} &+ \sum_{\alpha}^{1,\dots,m} a^{r_1\cdots r_{\alpha-1} j r_{\alpha+1}\cdots r_m}_{s_1\cdots s_p} L^{r_\alpha}_{ji} \\ &- \sum_{\beta}^{1,\dots,p} a^{r_1\cdots r_m}_{s_1\cdots s_{\beta-1} k s_{\beta+1}\cdots s_p} L^k_{s_\beta i} \end{aligned} \tag{3.2}$$

are the components of a tensor of order $m+p+1$, the first covariant derivative of the given tensor. As a consequence of this definition we have

* 1926, 1, p. 19.

† Cf. *Schouten*, 1924, 1, p. 83.

The first covariant derivative with respect to the L's of the tensor δ^i_j is zero.

If in (3.2) the L's are replaced by the corresponding Christoffel symbols (1.6) of the second kind, we obtain the formulas for covariant differentiation with respect to the fundamental tensor of a Riemannian geometry.* As in the latter we can establish the theorem:

Covariant differentiation of the sum, difference, outer and inner product of tensors obeys the same rules as ordinary differentiation.

In order that this theorem may hold for the case of an invariant obtained by the inner multiplication of a contravariant and a covariant vector, it is necessary that we define the first covariant derivative of an invariant (or scalar) to be its ordinary derivative.

Since the Christoffel symbols $\begin{Bmatrix} i \\ j\,k \end{Bmatrix}$ are symmetric in j and k, equations (3.2) are not the only generalization of covariant differentiation in Riemannian geometry. Thus if in (3.2) we replace $L^{r_\alpha}_{ji}$ and $L^k_{s_\beta i}$ by $L^{r_\alpha}_{ij}$ and $L^k_{is_\beta}$, we again obtain components of a tensor, as follows from the following considerations. When we put

$$L^i_{jk} - L^i_{kj} = 2\,\Omega^i_{jk}, \tag{3.3}$$

we have from (2.1) that Ω^i_{jk} are the components of a tensor. Consequently the differences between the quantities defined by (3.2) and those obtained by the change described above are the components of a tensor.

Still other definitions of covariant differentiation are possible. Thus recently Einstein† was led to the consideration of the equations

$$\frac{\partial a_{ij}}{\partial x^k} - a_{ih}\,L^h_{kj} - a_{hj}\,L^h_{ik} = 0.$$

From (3.2) and (3.3) it follows that the left-hand members of these equations are the components of a tensor. However,

* 1926, 1, p. 28.

† 1925, 11.

if $a_{ij} = \lambda_i \mu_j$, the above theorem as regards products does not hold,* if we define the left-hand member to be the covariant derivative of a_{ij}.

When dealing with asymmetric connections, we shall adhere to (3.2) as the formula for covariant differentiation. Several subscripts preceded by a solidus (|) indicate repeated covariant differentiation with respect to the L's.

4. Generalized identities of Ricci. If θ is an invariant, its second covariant derivative is given by

$$\theta_{\,ij} = \frac{\partial^2 \theta}{\partial x^i\, \partial x^j} - L^k_{ij}\, \theta_{|k}.$$

From this expression it follows that

$$\theta_{\,ij} - \theta_{\,ji} = -2\,\theta_{|k}\, \Omega^k_{ij}, \tag{4.1}$$

where Ω^k_{ij} are defined by (3.3); we recall also that they are the components of a tensor.

Proceeding in like manner with a contravariant vector λ^i, a covariant vector λ_i and a covariant tensor a_{ij}, we obtain respectively

$$\lambda^i_{\;|jk} - \lambda^i_{\;|kj} = -\lambda^h L^i_{hjk} - 2\,\lambda^i_{\;|h}\, \Omega^h_{jk}, \tag{4.2}$$

$$\lambda_{i|jk} - \lambda_{i|kj} = \lambda_h L^h_{ijk} - 2\,\lambda_{i|h}\, \Omega^h_{jk}, \tag{4.3}$$

$$a_{ij|kl} - a_{ij|lk} = a_{hj} L^h_{ikl} + a_{ih} L^h_{jkl} - 2\,a_{ij|h}\, \Omega^h_{kl}. \tag{4.4}$$

And in general we have

$$\begin{aligned} a^{r_1\cdots r_m}_{s_1\cdots s_p|kl} - a^{r_1\cdots r_m}_{s_1\cdots s_p|lk} = \sum_{\alpha}^{1,\cdots,p} a^{r_1\cdots r_m}_{s_1\cdots s_{\alpha-1} h s_{\alpha+1}\cdots s_p} L^h_{s_\alpha kl} \\ - \sum_{\beta}^{1,\cdots,m} a^{r_1\cdots r_{\beta-1} h r_{\beta+1}\cdots r_m}_{s_1\cdots s_p} L^{r_\beta}_{hkl} - 2\,a^{r_1\cdots r_m}_{s_1\cdots s_p|h}\,\Omega^h_{kl}. \end{aligned} \tag{4.5}$$

The foregoing identities are generalizations of the Ricci identities of Riemannian geometry.† When covariant differentiation is used, it is advantageous to use (4.5) in place of the

* Cf. *J. M. Thomas*, 1926, 13, p. 189.

† 1926, 1, p. 30; cf. *Schouten*, 1924, 1, p. 85.

customary conditions of integrability of ordinary differentiation, namely $\frac{\partial}{\partial x^j}\left(\frac{\partial \theta}{\partial x^i}\right) = \frac{\partial}{\partial x^i}\left(\frac{\partial \theta}{\partial x^j}\right)$, which are used in fact in the derivation of the *generalized identities of Ricci*.

5. Other fundamental tensors. If we put

$$\Gamma^i_{jk} = \frac{1}{2}(L^i_{jk} + L^i_{kj}), \tag{5.1}$$

it follows from these equations and (3.3) that

$$L^i_{jk} = \Gamma^i_{jk} + \Omega^i_{jk}. \tag{5.2}$$

Thus Γ^i_{jk} and Ω^i_{jk} are the symmetric and skew-symmetric parts respectively of L^i_{jk}. Substituting these expressions in (2.6), we obtain

$$L^i_{jkl} = B^i_{jkl} + \Omega^i_{jkl}, \tag{5.3}$$

where

$$B^i_{jkl} = \frac{\partial \Gamma^i_{jl}}{\partial x^k} - \frac{\partial \Gamma^i_{jk}}{\partial x^l} + \Gamma^h_{jl}\Gamma^i_{hk} - \Gamma^h_{jk}\Gamma^i_{hl} \tag{5.4}$$

and

$$\Omega^i_{jkl} = \Omega^i_{jl|k} - \Omega^i_{jk|l} + \Omega^i_{hl}\,\Omega^h_{jk} - \Omega^i_{hk}\,\Omega^h_{jl} - 2\,\Omega^i_{jh}\,\Omega^h_{kl}. \tag{5.5}$$

From equations (2.1) and (5.2) we have

$$\frac{\partial^2 x^i}{\partial x'^\beta \partial x'^\gamma} + \Gamma^i_{jk}\frac{\partial x^j \partial x^k}{\partial x'^\beta \partial x'^\gamma} = \Gamma'^\alpha_{\beta\gamma}\frac{\partial x^i}{\partial x'^\alpha}. \tag{5.6}$$

Since these equations are of the form (2.1), it follows that B^i_{jkl} are the components of a tensor. This is evident also from (5.3), since Ω^i_{jkl} are the components of a tensor.

From (5.3), (5.4) and (5.5) it follows that the tensors L^i_{jkl}, B^i_{jkl} and Ω^i_{jkl} are skew-symmetric in the indices k and l.

If B_{jk} denotes the contracted tensor B^i_{jki} and b_{jk} and β_{jk} denote respectively the symmetric and skew-symmetric parts of B_{jk}, we have from (5.4)

$$b_{jk} = \frac{1}{2}\left(\frac{\partial \Gamma^h_{hj}}{\partial x^k} + \frac{\partial \Gamma^h_{hk}}{\partial x^j}\right) - \frac{\partial \Gamma^h_{jk}}{\partial x^h} + \Gamma^h_{ji}\Gamma^i_{hk} - \Gamma^h_{jk}\Gamma^i_{hi}, \tag{5.7}$$

$$\beta_{jk} = \frac{1}{2}\left(\frac{\partial \Gamma^h_{hj}}{\partial x^k} - \frac{\partial \Gamma^h_{hk}}{\partial x^j}\right). \tag{5.8}$$

We shall show that β_{jk} is the curl of a vector. In fact, let g_{ij} be any symmetric tensor of the second order and form the Christoffel symbols of the second kind; if we put

$$\Gamma^i_{jk} = \begin{Bmatrix} i \\ j\,k \end{Bmatrix} + a^i_{jk}, \tag{5.9}$$

it follows from (1.5) and (5.6) that a^i_{jk} are the components of a tensor symmetric in the indices j and k. Since

$$\begin{Bmatrix} i \\ i\,j \end{Bmatrix} = \frac{\partial \log \sqrt{g}}{\partial x^j}\,*,$$

it follows from (5.8) and (5.9) that

$$\beta_{jk} = \frac{1}{2}\left(\frac{\partial a_j}{\partial x^k} - \frac{\partial a_k}{\partial x^j}\right), \quad a_j \equiv a^i_{ij} = \Gamma^i_{ij} - \begin{Bmatrix} i \\ i\,j \end{Bmatrix}. \tag{5.10}$$

If in place of taking the tensor g_{ij} we had taken any other tensor $\bar{g}_{ij}$, the function $g/\bar{g}$ is a scalar, and consequently a_j in (5.10) would have been replaced by a_j plus a gradient.

From (5.4) and (5.8) we obtain

$$S_{kl} \equiv B^i_{ikl} = 2\,\beta_{lk}. \tag{5.11}$$

If we put

$$\Omega_j = \Omega^i_{ij} = -\Omega^i_{ji}, \tag{5.12}$$

we have from (5.5) by contraction for i and l and for i and j respectively

$$\Omega_{jk} \equiv \Omega^i_{jki} = -(\Omega_{j/k} + \Omega^i_{jk/i} + \Omega_h\,\Omega^h_{jk} + \Omega^i_{jh}\,\Omega^h_{ki}) \tag{5.13}$$

and

$$\Phi_{kl} \equiv \Omega^i_{ikl} = \frac{\partial \Omega_l}{\partial x^k} - \frac{\partial \Omega_k}{\partial x^l}. \tag{5.14}$$

As a consequence of (5.3), (5.10), (5.11) and (5.14) we have:

The skew-symmetric tensor L^i_{ikl} is the curl of the vector $-(a_i + \Omega_i)$, where a_i is determined to within an additive arbitrary gradient.

* 1926, 1, p. 18.

When the expressions (5.9) are substituted in (5.4) and we denote by R^i_{jkl} the components of the Riemannian curvature tensor for the tensor g_{ij}, we have

$$(5.15)\qquad B^i_{jkl} = R^i_{jkl} + a^i_{jl;k} - a^i_{jk;l} + a^h_{jl}\, a^i_{hk} - a^h_{jk}\, a^i_{hl},$$

where a semi-colon followed by an index indicates covariant differentiation with respect to the g's. Contracting for i and l and for i and j, we have

$$(5.16)\qquad B_{jk} = R_{jk} + a_{j;k} - a^i_{jk;i} + a^h_{ji}\, a^i_{hk} - a^h_{jk}\, a_h$$

and (5.10), in consequence of (5.11). From (5.16) it follows that the symmetric part of B_{jk} is

$$(5.17)\quad b_{jk} = R_{jk} + \frac{1}{2}(a_{j;k} + a_{k;j}) - a^i_{jk;i} + a^h_{ji}\, a^i_{hk} - a^h_{jk}\, a_h.$$

If the symmetric tensor b_{ij}, defined by (5.7), satisfies the condition that the determinant $|b_{ij}|$ is not identically zero, it may be made to play a role for the manifold analogous in some respects to that of the fundamental tensor in Riemannian geometry. It is the tensor which would naturally be used for the tensor g_{ij} in the above equations to give determinateness to these equations.*

From (5.8), (5.10) and (5.11) we have

$$S_{ij} = \frac{\partial \Gamma^h_{hj}}{\partial x^i} - \frac{\partial \Gamma^h_{hi}}{\partial x^j} = \frac{\partial a_j}{\partial x^i} - \frac{\partial a_i}{\partial x^j},$$

from which it follows that a function g is defined by the equations

$$(5.18)\qquad \frac{\partial \log \sqrt{g}}{\partial x^i} = \Gamma^h_{hi} - a_i.$$

From the relation (5.6) it follows that between g and the corresponding function g' in another coördinate system x'^i we have the relation

$$\sqrt{g} = \sqrt{g'}\left|\frac{\partial x'^i}{\partial x^j}\right|$$

* Cf. *Eisenhart,* 1923, 4, p. 373.

Consequently we have the invariant integral

$$I = \int \sqrt{g}\, dx^1 \cdots dx^n = \int \sqrt{g'}\, dx'^1 \cdots dx'^n .$$

If in (5.18) we replace a_i by $a_i + \dfrac{\partial \log \theta}{\partial x^i}$, where θ is an arbitrary scalar, then $\sqrt{g}$ is replaced by $\sqrt{g}\,\theta$. Hence for a given symmetric connection there is no uniquely defined fundamental integral like the volume integral of a Riemannian space. If, however, the tensor S_{ij} is zero for the connection, the function g defined by (5.18) with $a_i = 0$ is uniquely defined and thus we have a volume integral for the space* which is analogous to that of a Riemannian space.†

6. Covariant differentiation with respect to the Γ's. Since the Γ's satisfy (5.6), which are of the form (2.1), it follows from (3.2) that the quantities

$$\begin{aligned}(6.1)\qquad a^{r_1\cdots r_m}_{s_1\cdots s_p, i} = \frac{\partial a^{r_1\cdots r_m}_{s_1\cdots s_p}}{\partial x^i} &+ \sum_{\alpha}^{1,\cdots,m} a^{r_1\cdots r_{\alpha-1} j r_{\alpha+1}\cdots r_m}_{s_1\cdots s_p}\, \Gamma^{r_\alpha}_{ji} \\ &- \sum_{\beta}^{1,\cdots,p} a^{r_1\cdots r_m}_{s_1\cdots s_{\beta-1} k s_{\beta+1}\cdots s_p}\, \Gamma^{k}_{s_\beta i}\end{aligned}$$

are the components of a tensor. This may be seen also by substituting the expressions (5.2) in (3.2) and observing that the differences between the resulting expressions and (6.1) are components of a tensor. The process defined by (6.1) we call *covariant differentiation with respect to the Γ's* and use a comma followed by indices to denote this type of covariant differentiation.‡

In terms of covariant differentiation with respect to the Γ's equations (5.5) become

$$(6.2)\qquad \Omega^i_{jkl} = \Omega^i_{jl,k} - \Omega^i_{jk,l} + \Omega^h_{jl}\,\Omega^i_{hk} - \Omega^h_{jk}\,\Omega^i_{hl}.$$

* Cf. *Veblen*, 1923, 8; *Eisenhart*, 1923, 9.

† 1926, p. 18.

‡ Cf. 1926, I, p. 28.

If θ is an invariant, we have

$$\theta_{,ij} - \theta_{,ji} = 0. \tag{6.3}$$

Also we have the following generalized identities of Ricci:

$$\begin{aligned} a^{r_1\cdots r_m}_{s_1\cdots s_p,kl} - a^{r_1\cdots r_m}_{s_1\cdots s_p,lk} &= \sum_{\alpha}^{1,\cdots,p} a^{r_1\cdots r_m}_{s_1\cdots s_{\alpha-1}hs_{\alpha+1}\cdots s_p} B^h_{s_\alpha kl} \\ &\quad - \sum_{\beta}^{1,\cdots m} a^{r_1\cdots r_{\beta-1}hr_{\beta+1}\cdots r_m}_{s_1\cdots s_p} B^{r_\beta}_{hkl}. \end{aligned} \tag{6.4}$$

7. Parallelism. Paths. In a general manifold there is no *a priori* basis for the comparison of vectors at different points. For a Riemannian manifold parallelism of vectors, as defined by Levi-Civita,* serves as a basis for such comparison. This definition may be generalized for a connected manifold. We say that a *curve* is the locus of points for which the coördinates x^i are functions of a parameter t. Let C be any curve and consider the system of differential equations

$$\frac{d\lambda^i}{dt} + L^i_{jk}\lambda^j \frac{dx^k}{dt} = 0, \tag{7.1}$$

where the x's in the L's are replaced by the functions of t for C. A solution of these equations, that is a set of functions $\lambda^1, \cdots, \lambda^n$ satisfying them, is determined by arbitrary values of the λ's for a given value of t, in accordance with the theory of differential equations.

Consider such a solution. Since the λ's are functions of t and likewise the x's, the λ's are expressible, in many ways, as functions of the x's. Assume that the λ's considered as functions of the x's are substituted in (7.1) and that the resulting equations are multiplied by $\dfrac{\partial x'^\alpha}{\partial x^i}$ and i is summed, x'^α being the coördinates of any other system for the space. By means of equations obtained from (2.1) by interchanging the x's and x''s, the resulting equation is reducible to

$$\frac{d\lambda'^\alpha}{dt} + L'^\alpha_{\beta\gamma}\lambda'^\beta \frac{dx'^\gamma}{dt} = 0,$$

* 1917, 1.

where

$$\lambda'^{\alpha} = \lambda^i \frac{\partial x'^{\alpha}}{\partial x^i}$$

Consequently a set of functions λ^i satisfying (7.1) are for each value of t the components of a contravariant vector. We say that they are *parallel* to one another *with respect to the curve*, and that any one of them may be obtained from any other by a *parallel displacement* of the latter along the curve. From the above remarks it follows that a family of vectors exists parallel to any given vector at a point of C. Since parallelism has thus been defined in terms of the connection, we say that the connection is *affine* and that the L's are the coefficients of affine connection.

Two vectors at a point are said to have the *same direction*, if corresponding components are proportional. Accordingly, if a set of functions λ^i satisfy equations (7.1), the vectors of components

$$\bar{\lambda}^i = \varphi \lambda^i, \tag{7.2}$$

where φ is any function of t, should be interpreted as parallel with respect to the given curve C. From (7.1) and (7.2) we have

$$\frac{d\bar{\lambda}^i}{dt} + L^i_{jk} \bar{\lambda}^j \frac{dx^k}{dt} = \bar{\lambda}^i f(t), \tag{7.3}$$

where

$$f(t) = \frac{d \log \varphi}{dt} \tag{7.4}$$

Conversely, if we have any set of functions $\bar{\lambda}^i$ of t which satisfy (7.3), they are the components of a family of contravariant vectors parallel with respect to C; and by means of (7.2) and (7.4) we find the vectors λ^i satisfying (7.1).

From (7.3) we have, on eliminating $f(t)$ and omitting the bars,

$$\lambda^h \left(\frac{d\lambda^i}{dt} + L^i_{jk} \lambda^j \frac{dx^k}{dt}\right) - \lambda^i \left(\frac{d\lambda^h}{dt} + L^h_{jk} \lambda^j \frac{dx^k}{dt}\right) = 0 \tag{7.5}$$

as the conditions of parellelism which hold for (7.2) whatever be φ.

As a particular example of the foregoing we consider the curves whose tangents are parallel with respect to the curves. From (7.5) it follows that the equations of these curves are

$$\begin{aligned}(7.6)\qquad &\frac{dx^j}{dt}\left(\frac{d^2x^i}{dt^2}+\Gamma^i_{kl}\frac{dx^k}{dt}\frac{dx^l}{dt}\right)\\ &-\frac{dx^i}{dt}\left(\frac{d^2x^j}{dt^2}+\Gamma^j_{kl}\frac{dx^k}{dt}\frac{dx^l}{dt}\right)=0,\end{aligned}$$

and that, conversely, any curve defined by these equations possesses the above property. We call these curves the *paths* of the manifold. They are an evident generalization of the geodesics of a Riemannian manifold.*

From the form of (7.6) it is evident that all connected spaces for which the Γ's are the same but Ω^i_{jk} are arbitrary have the same paths. Later (§ 12) it will be shown that this is not a necessary condition.

8. A theorem on partial differential equations. Consider a system of partial differential equations

$$(8.1)\qquad \frac{\partial\theta^\alpha}{\partial x^i}=\psi^\alpha_i(\theta,x)\quad(\alpha=1,\cdots,M;\ i=1,\cdots,n),$$

where the ψ's are functions of the θ's and x's. The conditions of integrability of these equations are

$$(8.2)\qquad \frac{\partial\psi^\alpha_i}{\partial x^j}+\frac{\partial\psi^\alpha_i}{\partial\theta^\beta}\frac{\partial\theta^\beta}{\partial x^j}=\frac{\partial\psi^\alpha_j}{\partial x^i}+\frac{\partial\psi^\alpha_j}{\partial\theta^\gamma}\frac{\partial\theta^\gamma}{\partial x^i},$$

where β and γ are summed from 1 to M. If these equations are satisfied identically, the system (8.1) is completely integrable and the general solution involves M arbitrary constants. For in this case we can obtain developments in powers of the x's, with constant coefficients, which satisfy (8.1), the coefficients being determined by the initial values of the θ's.

* 1926, 1, p. 50.

If equations (8.2) are not satisfied identically, we have a set F_1 of equations, which establish conditions upon the θ's as functions of the x's. If we differentiate each of these equations with respect to the x's and substitute for $\frac{\partial \theta^\alpha}{\partial x^i}$ from (8.1), either the resulting equations are a consequence of the set F_1 or we get a new set F_2. Proceeding in this way we get a sequence of sets, $F_1, F_2, \cdots$, of equations, which must be compatible, if equations (8.1) are to have a solution. If one of these sets is not a consequence of the preceding sets, it introduces at least one additional condition. Consequently, if the equations (8.1) are to admit a solution, there must be a positive integer N such that the equations of the $(N+1)$th set are satisfied because of the equations of the preceding N sets; otherwise we should obtain more than M independent equations which would imply a relation between the x's. Moreover, from this argument it follows that $N \leqq M$.

Conversely, suppose that there is a number N such that the equations of the sets

$$F_1, \cdots, F_N, \tag{8.3}$$

are compatible and each set introduces one or more conditions independent of the conditions imposed by the equations of the other sets, and that all of the equations of the set

$$F_{N+1} \tag{8.4}$$

are satisfied identically because of the equations of the sets (8.3). Assume that there are p $(< M)$ independent conditions imposed by (8.3), say $G_\nu(\theta, x) = 0$. Since the jacobian matrix $\left\| \frac{\partial G_\nu}{\partial \theta^\alpha} \right\|$ is of rank p, these equations may be regarded as solved for p of the θ's in terms of the remaining θ's and the x's, and the equations are then of the form (by suitable numbering)

$$\theta^\sigma - \varphi^\sigma(\theta^{p+1}, \cdots, \theta^M, x) = 0 \quad (\sigma = 1, \cdots, p). \tag{8.5}$$

From these equations we have by differentiation

$$\frac{\partial \theta^\sigma}{\partial x^i} - \frac{\partial \varphi^\sigma}{\partial \theta^\nu}\frac{\partial \theta^\nu}{\partial x^i} - \frac{\partial \varphi^\sigma}{\partial x^i} = 0 \quad (\nu = p+1, \cdots, M).$$

Replacing $\dfrac{\partial \theta^\sigma}{\partial x^i}$ by means of (8.1), we have

$$\psi_i^\sigma - \frac{\partial \varphi^\sigma}{\partial \theta^\nu}\psi_i^\nu - \frac{\partial \varphi^\sigma}{\partial x^i} = 0,$$

which equations are satisfied because of the sets (8.3) and (8.4), as follows from the method of obtaining the latter. Accordingly we have by subtraction

$$\text{(8.6)} \qquad \frac{\partial \theta^\sigma}{\partial x^i} - \psi_i^\sigma - \frac{\partial \varphi^\sigma}{\partial \theta^\nu}\left(\frac{\partial \theta^\nu}{\partial x^i} - \psi_i^\nu\right) = 0.$$

From these equations it follows that, if the functions $\theta^{p+1}, \cdots, \theta^M$ are chosen to satisfy the equations

$$\text{(8.7)} \qquad \frac{\partial \theta^\nu}{\partial x^i} = \bar{\psi}_i^\nu (\theta^{p+1}, \cdots, \theta^M, x),$$

where $\bar{\psi}_i^\nu$ is obtained from ψ_i^ν on replacing θ^σ $(\sigma = 1, \cdots, p)$ by their expressions (8.5), then equations (8.1) for $\alpha = 1, \cdots, p$ are satisfied by the values (8.5). Since the equations of the set F_1 are satisfied identically because of (8.5), it follows that equations (8.7) are completely integrable; for, the equations arising from expressing their conditions of integrability are in the set F_1. Consequently there is a solution in this case and it involves $M-p$ arbitrary constants.

When $p = M$, we have in place of (8.5) $\theta^\alpha = \varphi^\alpha(x)$ and in place of (8.6) that the functions θ^α satisfy (8.1). In this case there are no constants of integration. Hence we have:

In order that a system of equations (8.1) *admit a solution, it is necessary and sufficient that there exist a positive integer* $N (\leq M)$ *such that the equations of the sets* $F_1, \cdots, F_N$ *are compatible for all values of the x's in a domain, and that* *the equations of the set F_{N+1} are satisfied because of the former sets; if p is the number of independent equations in the first N sets, the solution involves $M-p$ arbitrary constants.**

It is evident from the above considerations that when an integer N exists such that the conditions of the theorem are satisfied, they are satisfied also for any integer larger than N. However, it is understood in the theorem and in the various applications of it that N is the least integer for which the conditions are satisfied.

The above theorem can be applied also to the case when there are certain functional relations between the θ's and x's which must be satisfied in addition to the differential equations (8.1). In this case we denote by F_0 this set of conditions, and include in the set F_1 of the theorem also such conditions as arise from F_0 by differentiation and substitution from (8.1). Then the theorem proceeds as above with the understanding that the sets $F_0, F_1, \cdots, F_N$ shall be compatible, and that the set F_{N+1} shall be satisfied because of the former.

In certain cases (cf. § 36) the equations of the set F_1 consist of two sets F_1' and F_1'', such that, if F_2' and F_2'' are those which follow from F_1' and F_1'' respectively, then the set F_1'' is a consequence of F_2'. In this case equations F_2'' are a consequence of F_3' and so on. Hence we have that all the solutions of $F_1', \cdots, F_{N+1}'$ satisfy the set F_{N+2}'. Accordingly in applying the theorem we have only to consider the sequence $F_1', \cdots, F_r', \cdots$.

When the functions ψ_i^α in (8.1) are linear and homogeneous in the θ's, the same is true of the equations of the sets $F_1, F_2, \cdots$; moreover p is at most equal to $M-1$. From algebraic considerations it follows that the conditions of the problem are that there exist a positive integer N such that

* This theorem for the case $M = p$ was used by *Christoffel*, 1869, 1, p. 60 in the solution of a certain problem (cf. § 28) and was used for the general case in the same problem by *Wright*, 1908, 1, pp. 16, 17; cf. also, *Bianchi*, 1918, 1, pp. 9–13; *Levi-Civita*, 1925, 5, pp. 40–43 and *Veblen* and *J. M. Thomas*, 1926, 6, pp. 288–290.

the rank of the matrix of the sets $F_1, \cdots, F_N$ is $M-p(p \geqq 1)$ and that this is also the rank of the matrix of $F_1, \cdots, F_{N+1}$. When these conditions are satisfied, the solution of the problem reduces to the integration of a completely integrable set of equations (8.7), in which now the ψ's are linear and homogeneous. Consequently any solution is expressible as a linear function with constant coefficients of p particular solutions, and such an expression with arbitrary constant coefficients is a solution. Most of the applications of this theorem which we shall make are to equations of this linear type. Moreover, these equations are of the form in which the θ's are components of a tensor and in place of their derivatives we have first covariant derivatives.

9. Fields of parallel contravariant vectors. When we have any contravariant vector-field of components λ^i, the vectors of the field at points of a curve C are parallel, if

$$\frac{dx^k}{dt}(\lambda^h \lambda^i_{|k} - \lambda^i \lambda^h_{|k}) = 0, \tag{9.1}$$

as follows from (7.5). In order that these equations be satisfied for the vectors of the field along any curve of the space it is necessary that

$$\lambda^h \lambda^i_{|k} - \lambda^i \lambda^h_{|k} = 0,$$

from which it follows that

$$\lambda^i_{|k} = \lambda^i \mu_k, \tag{9.2}$$

where μ_k is a covariant vector. When μ_k is not a gradient, the function $\mu_k \dfrac{dx^k}{dt}$ depends upon the curve, so that if the vector λ^i at a point P is subjected to parallel displacement around a closed circuit the resulting vector at P will depend upon the path; this is shown in § 10. This will not be the case if μ_k is a gradient, in which case (9.2) may be written

$$\lambda^i_{|k} = \lambda^i \frac{\partial \log \varphi}{\partial x^k}. \tag{9.3}$$

A field of vectors satisfying equations (9.3) is said to be a *parallel field.* If we change the components replacing λ^i by $\lambda^i \varphi$, the new components satisfy the equations

$$(9.4) \qquad \lambda^i{}_{|k} = \frac{\partial \lambda^i}{\partial x^k} + L^i_{jk} \lambda^j = 0.$$

If λ^i are the components of a parallel field they define a congruence of curves along any one of which it is possible to choose a parameter t so that $\lambda^i = \frac{d x^i}{d t}$. Then from (9.3) or (9.4) and (7.6) we have:

The curves of a congruence of curves determined by a field of parallel vectors are paths.

From (4.2) we have that the conditions of integrability of equations (9.4) are

$$(9.5) \qquad \lambda^h L^i_{hjk} = 0.$$

When

$$(9.6) \qquad L^i_{hjk} = 0,$$

equations (9.4) are completely integrable, that is, a solution is uniquely determined by arbitrary values of λ^i at a given point. Hence we have:

A necessary and suffiicient condition that there exist a field of contravariant vectors parallel to an arbitrary vector is that (9.6) *be satisfied.*

From equations (9.5) we have also:

A necessary and sufficient condition that a V_n *admit* n *linearly independent fields of parallel contravariant vectors is that the curvatnre tensor* L^i_{jkl} *be a zero tensor.*

If equations (9.6) are not satisfied, on differentiating (9.5) covariantly, we have in consequence of (9.4)

$$(9.7) \qquad \lambda^h L^i_{hjk\,m_1} = 0.$$

Proceeding in like manner with these equations, we obtain the sequence of equations

$$\lambda^h L^i_{hjk\,m_1 m_2} = 0,$$

$$\cdots\cdots\cdots$$

$$\lambda^h L^i_{hjk\,m_1 m_2 \cdots m_q} = 0, \tag{9.8}$$

$$\cdots\cdots\cdots$$

$$\cdots\cdots\cdots$$

Equations (9.4) are of the form (8.1). Hence in consequence of the results of § 8 we have:

A necessary and sufficient condition for the existence of one or more fields of parallel contravariant vectors is that there exist a positive integer N such that the first N sets of equations (9.5), (9.7) *and* (9.8) *admit $r\ (\geqq 1)$ fundamental sets of solutions, which satisfy the $(N+1)$th set of equations; if these conditions are satisfied, there are r linearly independent fields of parallel vectors and any linear combination, with constant coefficients, of these vectors is a parallel field.**

Having thus obtained the conditions for one or more fields of parallel contravariant vectors in invariantive form, we shall show how all such fields may be obtained by making a suitable choice of coördinates.

Suppose we have r fields of parallel vectors of components $\lambda^i_{(\alpha)}$, where α, for $\alpha = 1, \cdots, r$, denotes the vector and i the component; we use the notation that an index in parentheses indicates an entity, one without parentheses a component. In another coördinate system x'^i we have

$$\lambda'^j_{(\alpha)} = \lambda^i_{(\alpha)} \frac{\partial x'^j}{\partial x^i}. \tag{9.9}$$

Consider the system of differential equations

$$X_\alpha(\theta) \equiv \lambda^i_{(\alpha)} \frac{\partial \theta}{\partial x^i} = 0. \tag{9.10}$$

Since by hypothesis the functions $\lambda^i_{(\alpha)}$ satisfy (9.4), the Poisson operator applied to equations (9.10) gives

* This theorem for the case $\Omega^i_{jk} = 0$ was established by *Veblen* and *T. Y. Thomas*, 1923, 1, p. 590.

$$(9.11) \qquad (X_\alpha X_\beta - X_\beta X_\alpha)\,\theta = 2\,\lambda^i_{(\alpha)}\,\lambda^j_{(\beta)}\,\Omega^k_{ij}\,\frac{\partial\theta}{\partial x^k}.^*$$

We consider first the case when $\Omega^k_{ij} = 0$, that is, when equations (9.4.) become

$$(9.12) \qquad \lambda^i_{,k} = 0.$$

In this case equations (9.10) form a complete system, and thus there are $n - r$ independent solutions $\theta^\sigma(x^1, \cdots, x^n)$ for $\sigma = r+1, \cdots, n$. If we omit any one of equations (9.10), the remaining ones form a complete system and admit in addition to the above another independent solution. In this way we get r other functions $\theta^\alpha(x^1, \cdots, x^n)$ for $\alpha = 1, \cdots, r$, θ^α being the additional solution when $X_\alpha(\theta) = 0$ is omitted. If we put

$$(9.13) \qquad x'^i = \theta^i(x^1, \cdots, x^n),$$

from (9.9) and (9.10) it follows that in the coördinate system x'^i, the components $\lambda'^i_{(\alpha)}$ are zero unless $i = \alpha$.

Suppose now that equations (9.12) are expressed in this coördinate system, which we call x^i; then the components of the r vectors are of the form

$$(9.14) \qquad \lambda^i_{(\alpha)} = \delta^i_\alpha\,\psi_\alpha,$$

α not being summed. From (9.12) we have

$$(9.15) \qquad \Gamma^i_{\alpha k} = -\,\delta^i_\alpha\,\frac{\partial \log \psi_\alpha}{\partial x^k}.$$

Since the Γ's must be symmetric in the lower indices, it is necessary that ψ_α be a function at most of $x^\alpha, x^{r+1}, \cdots, x^n$. Hence we have the theorem:

The most general space with a symmetric connection admitting r fields of parallel contravariant vectors is obtained by choosing arbitrarily the coefficients $\Gamma^i_{\sigma\tau}$ for σ and τ equal

* Cf. *Goursat*, 1891, 1, p. 52.

to $r+1, \cdots, n$, *and for the others expressions of the form* (9.15). *where* ψ_α *is a function of* $x^\alpha, x^{r+1}, \cdots, x^n$.*

When the connection is asymmetric, the quantities (9.14) satisfy (9.4) if

$$L^i_{\alpha k} = -\delta^i_\alpha \frac{\partial \log \psi_\alpha}{\partial x^k} \qquad (\alpha = 1, \cdots, r), \tag{9.16}$$

where α is not summed. Hence we have:

A space with asymmetric connection admitting r fields of parallel contravariant vectors is defined by (9.16) *where* ψ_α *are any functions of the x's and the other L's are arbitrary.*

In particular, if $r = n$, it follows from the above results that the tensor L^i_{jkl} is zero. This is readily verified for the expressions (9.16).

If in equations (2.1) we replace $L'^\gamma_{\alpha\beta}$ by expressions of the form (9.16) for $\alpha, \beta, \gamma = 1, \cdots, n$, we have

$$\frac{\partial^2 x^i}{\partial x'^\alpha \partial x'^\beta} + L^i_{jk} \frac{\partial x^j}{\partial x'^\alpha} \frac{\partial x^k}{\partial x'^\beta} + \frac{\partial \log \psi_\alpha}{\partial x'^\beta} \frac{\partial x^i}{\partial x'^\alpha} = 0,$$

where α is not summed in the last term. Since by hypothesis L^i_{jkl} are zero, and $L'^\sigma_{\alpha\beta\gamma}$ are zero when ψ_α are arbitrary functions of the x's, it follows from § 2 that the above equations are completely integrable. Hence we have:

When the curvature tensor of a space with asymmetric connection is zero, a coördinate system exists for which the coefficients have the form (9.16), *the n functions* ψ_α *being arbitrary.*

If the ψ's are constants, the coefficients must be zero, which is possible only in case of a symmetric connection, as is evident from (2.1) if we take $L^i_{jk} = 0$. In this case we have as a corollary of the above theorem:

When the curvature tensor of a space with a symmetric connection is zero, a coördinate system exists for which all of the coefficients of the connection are zero.

10. Parallel displacement of a contravariant vector around an infinitesimal circuit. In order to consider

* *Eisenhart*, 1922, 1, p. 210.

the parallel displacement of a vector around an infinitesimal circuit, we consider a surface, that is a manifold of two dimensions, defined by equations $x^i = f^i(u, v)$, where the functions f and their derivatives to the third order exist and are continuous at a point P of the surface. We consider the circuit comprising the points $P(u, v)$, $Q(u+du, v)$, $R(u+du, v+dv)$, $S(u, v+dv)$ and P. We take a vector λ^i at P and find from (7.1) the components of the vector at Q parallel to it, then in the same way the vector at R parallel to this vector at Q and so on. The components of the resulting vectors are given by

$$
(10.1)\quad
\begin{aligned}
(\lambda^i)_Q &= (\lambda^i)_P + \left(\frac{d\lambda^i}{du}\right)_P du + \frac{1}{2}\left(\frac{d^2\lambda^i}{du^2}\right)_P du^2 + \cdots, \\
(\lambda^i)_R &= (\lambda^i)_Q + \left(\frac{d\lambda^i}{dv}\right)_Q dv + \frac{1}{2}\left(\frac{d^2\lambda^i}{dv^2}\right)_Q dv^2 + \cdots, \\
(\lambda^i)_S &= (\lambda^i)_R - \left(\frac{d\lambda^i}{du}\right)_R du + \frac{1}{2}\left(\frac{d^2\lambda^i}{du^2}\right)_R du^2 + \cdots, \\
(\bar{\lambda}^i)_P &= (\lambda^i)_S - \left(\frac{d\lambda^i}{dv}\right)_S dv + \frac{1}{2}\left(\frac{d^2\lambda^i}{dv^2}\right)_S dv^2 + \cdots,
\end{aligned}
$$

where the quantities such as $\left(\frac{d\lambda^i}{dv}\right)_Q$, $\left(\frac{d^2\lambda^i}{dv^2}\right)_Q$, and so forth are obtained from equations of the form (7.1). When all of the above equations are added, we obtain

$$
\begin{aligned}
\Delta(\lambda^i)_P \equiv (\bar{\lambda}^i)_P - (\lambda^i)_P &= \left[\left(\frac{d\lambda^i}{du}\right)_P - \left(\frac{d\lambda^i}{du}\right)_R\right] du \\
(10.2)\quad + \left[\left(\frac{d\lambda^i}{dv}\right)_Q - \left(\frac{d\lambda^i}{dv}\right)_S\right] dv &+ \frac{1}{2}\left[\left(\frac{d^2\lambda^i}{du^2}\right)_P + \left(\frac{d^2\lambda^i}{du^2}\right)_R\right] du^2 \\
+ \frac{1}{2}\left[\left(\frac{d^2\lambda^i}{dv^2}\right)_Q + \left(\frac{d^2\lambda^i}{dv^2}\right)_S\right] dv^2 &+ \cdots.
\end{aligned}
$$

At P we have

$$
(10.3)\quad
\begin{aligned}
\left(\frac{d\lambda^i}{du}\right)_P &= -\left(L^i_{jk}\lambda^j \frac{\partial x^k}{\partial u}\right)_P, \\
\left(\frac{d^2\lambda^i}{du^2}\right)_P &= -\left\{\lambda^j\left[\frac{\partial}{\partial u}\left(L^i_{jk}\frac{\partial x^k}{\partial u}\right) - L^i_{hk} L^h_{jl}\frac{\partial x^k}{\partial u}\frac{\partial x^l}{\partial u}\right]\right\}_P
\end{aligned}
$$

When the functions L^i_{jk} and $\dfrac{\partial x^k}{\partial v}$ at Q are replaced by their expansions about P and use is made of (7.1), we have

$$\begin{aligned}
\left(\frac{d\lambda^i}{dv}\right)_Q &= -\left(L^i_{jk}\,\lambda^j\frac{\partial x^k}{\partial v}\right)_P \\
&\quad -\left\{\lambda^j\left[\frac{\partial}{\partial u}\left(L^i_{jk}\frac{\partial x^k}{\partial v}\right)-L^i_{hk}\,L^h_{jl}\frac{\partial x^k}{\partial v}\frac{\partial x^l}{\partial u}\right]\right\}_P du+\cdots, \qquad (10.4)\\
\left(\frac{d^2\lambda^i}{dv^2}\right)_Q &= -\left\{\lambda^j\left[\frac{\partial}{\partial v}\left(L^i_{jk}\frac{\partial x^k}{\partial v}\right)-L^i_{hk}\,L^h_{jl}\frac{\partial x^k}{\partial v}\frac{\partial x^l}{\partial v}\right]\right\}_P+\cdots.
\end{aligned}$$

In like manner we obtain

$$\begin{aligned}
&\left(\frac{d\lambda^i}{du}\right)_P-\left(\frac{d\lambda^i}{du}\right)_R \\
&\quad= \left\{\lambda^j\left[\frac{\partial}{\partial u}\left(L^i_{jk}\,\frac{\partial x^k}{\partial u}\right)-L^i_{hk}\,L^h_{jl}\frac{\partial x^k}{\partial u}\,\frac{\partial x^l}{\partial u}\right]\right\}_P du \\
&\qquad+\left\{\lambda^j\left[\frac{\partial}{\partial v}\left(L^i_{jk}\,\frac{\partial x^k}{\partial u}\right)-L^i_{hk}\,L^h_{jl}\frac{\partial x^k}{\partial u}\,\frac{\partial x^l}{\partial v}\right]\right\}_P dv+\cdots, \\
&\left(\frac{d\lambda^i}{dv}\right)_Q-\left(\frac{d\lambda^i}{dv}\right)_S \\
&\quad= -\left\{\lambda^j\left[\frac{\partial}{\partial u}\left(L^i_{jk}\,\frac{\partial x^k}{\partial v}\right)-L^i_{hk}\,L^h_{jl}\frac{\partial x^k}{\partial v}\,\frac{\partial x^l}{\partial u}\right]\right\}_P du \\
&\qquad+\left\{\lambda^j\left[\frac{\partial}{\partial v}\left(L^i_{jk}\,\frac{\partial x^k}{\partial v}\right)-L^i_{hk}\,L^h_{jl}\frac{\partial x^k}{\partial v}\,\frac{\partial x^l}{\partial v}\right]\right\}_P dv+\cdots,
\end{aligned}$$

We remark also that the expressions for $\left(\dfrac{d^2\lambda^i}{du^2}\right)_R$ and $\left(\dfrac{d^2\lambda^i}{dv^2}\right)_S$ differ from those for $\left(\dfrac{d^2\lambda^i}{du^2}\right)_P$ and $\left(\dfrac{d^2\lambda^i}{dv^2}\right)_Q$ respectively, as given by (10.3) and (10.4), only in terms of the first and higher orders of the differentials. When these expressions are substituted in (10.2) we have

$$(10.5)\qquad \Delta(\lambda^i)_P = -\left(\lambda^j\,L^i_{jkl}\,\frac{\partial x^k}{\partial u}\,\frac{\partial x^l}{\partial v}\right)_P du\,dv+\cdots.$$

From the considerations of § 9 it follows that $\Delta(\lambda^i)_P=0$ when $L^i_{jkl}=0$. The same is true when λ^i belongs to a field of parallel vectors. But in the general case when a vector undergoes parallel displacement around an infinitesimal circuit the difference between its final and original position is of the second order and depends upon the value of the components L^i_{jkl} at the starting point.*

Let $\lambda^i_{(\alpha)}$ be the components of n independent contravariant vectors at P, where α, for $\alpha = 1, \cdots, n$, indicates the vector and i, for $i = 1, \cdots, n$, the component. If these vectors are displaced about an infinitesimal circuit and we denote by λ the determinant $|\lambda^i_{(\alpha)}|$, then from (10.5) we have

$$(10.6)\qquad \Delta(\lambda)_P = -\left(\lambda L^i_{ikl} \frac{\partial x^k}{\partial u} \frac{\partial x^l}{\partial v}\right)_P du\, dv + \cdots.$$

Hence for this variation to be of the third or higher order it is necessary that [cf. (2.6)]

$$L^i_{ikl} = \frac{\partial L^i_{il}}{\partial x^k} - \frac{\partial L^i_{ik}}{\partial x^l} = 0.\dagger$$

From these equations it follows that

$$(10.7)\qquad L^i_{ij} = \frac{\partial \log \varphi}{\partial x^j}.$$

In another coördinate system x'^i we have

$$L'^r_{rs} = \frac{\partial \log \varphi'}{\partial x'^s},$$

and we desire to find the relation between φ and φ'. From (2.1) we have

$$L'^r_{rs} = \frac{\partial}{\partial x'^s} \log \Delta + L^i_{ij} \frac{\partial x^j}{\partial x'^s},$$

where Δ is the jacobian $\left|\frac{\partial x^i}{\partial x'^r}\right|$. Consequently we have, to within a negligible constant factor,

* Cf. *Schouten*, 1924, 1, p. 84.
† Cf. *Schouten*, 1924, 1, p. 89.

(10.8) $$\varphi' = \varphi \Delta,$$

that is, φ is a scalar density.

If $\lambda'^i_{(\alpha)}$ denote the components of the vectors in the coördinates x'^i and λ' denotes the determinant $|\lambda'^i_{(\alpha)}|$, we have

(10.9) $$\lambda' = \frac{\lambda}{\Delta},$$

that is, λ is a relative invariant of weight -1. Accordingly $\lambda \varphi$ is an invariant (or scalar).

If now we take n linearly independent contravariant vectors parallel with respect to a curve C and let φ be any scalar density, we have

$$\frac{d}{dt} \lambda \varphi = \lambda \left(\frac{\partial \varphi}{\partial x^k} - \varphi L^i_{ik} \right) \frac{dx^k}{dt}.$$

Consequently if the invariant $\lambda \varphi$ so formed with respect to every curve in space is to be constant along the curve, it is necessary and sufficient that (10.7) hold, the function φ being thus determined.

In particular, if the connection is symmetric, we have in place of (10.7)

(10.10) $$\Gamma^i_{ij} = \frac{\partial \log \varphi}{\partial x^j}$$

Then from (5.8) and (5.11) we have that

(10.11) $$B^i_{ijk} = 0, \qquad B_{ij} = B_{ji}.$$

Conversely, when conditions (10.11) are satisfied, we have (10.10), as follows from (5.8) and (5.11). Hence we have:*

If for a symmetric connection the contracted tensor B_{ij} is symmetric, the magnitude of the determinant λ of n linearly independent contravariant vectors $\lambda^i_{(\alpha)}$ is unaltered to within terms of the third and higher order, when the vectors undergo parallel displacements about an infinitesimal circuit, and conversely.

* Cf. *Schouten*, 1924, 1, p. 90.

11. Pseudo-örthogonal contravariant and covariant vectors. Parallelism of covariant vectors. If μ_i are the components of any covariant vector, there are evidently $n-1$ linearly independent contravariant vectors $\lambda^i_{(\alpha)}$ such that

$$\mu_i \lambda^i_{(\alpha)} = 0. \qquad (\alpha = 1, \cdots, n-1). \tag{11.1}$$

We say that each of the vectors $\lambda^i_{(\alpha)}$ is *pseudo-örthogonal* to μ_i; in Riemannian geometry (11.1) is the definition of orthogonality, when μ_i are the covariant components of a contravariant vector μ^i.* Evidently any vector pseudo-örthogonal to μ_i is expressible in the form

$$\lambda^i = a^\alpha \lambda^i_{(\alpha)} \qquad (\alpha = 1, \cdots, n-1). \tag{11.2}$$

where the a's are invariants; here, and in similar cases later, α is supposed to be summed for its values $1, \cdots, n-1$.

Consider any curve C of the space and $n-1$ linearly independent families of contravariant vectors $\lambda^i_{(\alpha)}$ parallel with respect to C. From (7.3) it follows that we have

$$\frac{d\lambda^i_{(\alpha)}}{dt} + L^i_{jk}\lambda^j_{(\alpha)}\frac{dx^k}{dt} = \lambda^i_{(\alpha)} f_\alpha(t) \qquad (\alpha = 1, \cdots, n-1), \tag{11.3}$$

α being not summed in the right-hand member. The equations

$$\lambda^i_{(\alpha)}\mu_i = 0 \tag{11.4}$$

define, to within a common factor, the components μ_i of a family of covariant vectors pseudo-örthogonal to the given $\lambda^i_{(\alpha)}$. We say that these vectors μ_i are *parallel with respect to C.* Differentiating (11.4) with respect to t and making use of (11.3), we obtain

$$\lambda^i_{(\alpha)}\left(\frac{d\mu_i}{dt} - L^j_{ik}\mu_j\frac{dx^k}{dt}\right) = 0.$$

* 1926, 1, p. 38.

Comparing these equations with (11.4), we find

$$\frac{d\mu_i}{dt} - L^j_{ik}\mu_j \frac{dx^k}{dt} = \mu_i \varphi(t) \tag{11.5}$$

as a necessary condition of parallelism.

In order to show that it is a sufficient condition, we consider $n-1$ linearly independent covariant vectors $\overset{(\alpha)}{\mu_i}$* satisfying equations of the form (11.5), that is,

$$\frac{d\overset{(\alpha)}{\mu_i}}{dt} - L^j_{ik}\overset{(\alpha)}{\mu_j} \frac{dx^k}{dt} = \overset{(\alpha)}{\mu_i} \varphi^\alpha(t), \tag{11.6}$$

where α is not summed in the right-hand member. Then the equations

$$\lambda^i \overset{(\alpha)}{\mu_i} = 0$$

determine quantities λ^i, to within a common factor, which are the components of a contravariant vector pseudo-örthogonal to each of the $n-1$ vectors $\overset{(\alpha)}{\mu_i}$. Differentiating these equations, we find that λ^i satisfies equations of the form (11.3) and consequently defines a family of contravariant vectors parallel with respect to C. Suppose now that we have any family of vectors μ_i satisfying (11.5) and we associate with it $n-1$ vectors satisfying (11.6) all n being linearly independent. The vector μ_i and each set of $n-2$ of the set $\overset{(\alpha)}{\mu_i}$ determine a contravariant vector pseudo-örthogonal to μ_i. In this way we obtain $n-1$ linearly independent families of parallel contravariant vectors pseudo-örthogonal to the vectors μ_i. Hence we have:

Any family of covariant vectors whose components satisfy equations of the form (11.5) *are parallel with respect to the given curve, that is, they are pseudo-örthogonal to $n-1$ linearly independent families of contravariant vectors parallel with respect to the curve.*

* Here α, where $\alpha = 1, \cdots, n-1$, indicates the vector and i the component.

Incidentally we have:

Any $n-1$ linearly independent families of covariant vectors parallel with respect to a curve are pseudo-örthogonal to a family of contravariant vectors parallel with respect to the curve.

By processes analogous to those used in § 9 we have that when the equations

$$\frac{\partial \mu_i}{\partial x^k} - L^j_{ik}\,\mu_j = 0 \tag{11.7}$$

admit a solution, the vector-field μ_i is parallel. However, we cannot say that the existence of such a field is equivalent to the existence of $n-1$ linearly independent fields of parallel contravariant vectors. For on differentiating (11.4) covariantly and making use of (11.7) we have

$$\mu_i\,\lambda^i_{(\alpha)|k} = 0,$$

which are equivalent to

$$\lambda^i_{(\alpha)|k} = a^\beta_{\alpha k}\,\lambda^i_{(\beta)},$$

where for each value of α and β the a's are components of a covariant vector.

From (11.7) and (4.3) we have:

A necessary and sufficient condition for the existence of n linearly independent fields of parallel covariant vectors is that the curvature tensor be zero.

When the connection is symmetric, it follows from (11.7) that μ_i is the gradient of a function φ. Since in this case the covariant vector μ_i at any point in space is pseudo-örthogonal to every displacement in the hypersurface $\varphi = \text{const.}$ containing the point, we call it the *covariant pseudonormal* to the hypersurface. Hence we have:

When a space with symmetric connection admits a parallel field of covariant vectors, they are the covariant pseudonormals to a family of hypersurfaces.

12. Changes of connection which preserve parallelism. Let L^i_{jk} and $\overline{L}^i_{jk}$ be the coefficients of two different connections. We inquire whether it is possible that parallel directions along every curve in the space are the same for the two connections. To this end we make use of the equations of parallelism in the form (7.5). Subtracting these equations from the corresponding ones in the $\overline{L}$'s, we have

$$(\delta_r^h\, a_{jk}^i - \delta_r^i\, a_{jk}^h)\, \lambda^r\, \lambda^j \frac{dx^k}{dt} = 0,$$

where

$$a_{jk}^i = \overline{L}_{jk}^i - L_{jk}^i.$$

From (2.1) it is seen that a_{jk}^i are the components of a tensor. Since these equations must hold for any curve and for vectors parallel to any vector with respect to this curve, we must have

$$\delta_r^h\, a_{jk}^i + \delta_j^h\, a_{rk}^i - \delta_r^i\, a_{jk}^h - \delta_j^i\, a_{rk}^h = 0.$$

Contracting for h and r, we have

$$a_{jk}^i = 2\, \delta_j^i\, \psi_k,$$

where ψ_k is the vector defined by

$$2\, n\, \psi_k = a_{hk}^h.$$

Conversely, if we take

$$\overline{L}_{jk}^i = L_{jk}^i + 2\, \delta_j^i\, \psi_k, \tag{12.1}$$

where ψ_k is an arbitrary vector, the above conditions are satisfied. Hence we have:

Equations (12.1) *in which* ψ_i *is an arbitrary covariant vector defines the most general change of connection which preserves parallelism.**

From the form of equations (12.1) it is seen that both sets of coefficients cannot be symmetric in the subscripts. In § 14 we discuss the case where one set does possess this property. Hence we have:

It is not possible to have two symmetric connections with respect to which parallel directions along every curve in the space are the same for both connections.

When the condition for parallelism is written in the form (7.3), that is,

* Cf. *Friesecke*, 1925, 1, p. 106; also *J. M. Thomas*, 1926, 3, p. 662.

$$\frac{d\lambda^i}{dt} + L^i_{jk}\lambda^j \frac{dx^k}{dt} = \lambda^i f(t),$$

the function $f(t)$ for the connection (12.1) is given by

$$f = f + 2\psi_k \frac{dx^k}{dt}. \tag{12.2}$$

If we have a field of parallel vectors λ^i defined by

$$\frac{\partial \lambda^i}{\partial x^k} + \lambda^j \bar{L}^i_{jk} = 0,$$

then for the connection defined by (12.1) we have

$$\frac{\partial \lambda^i}{\partial x^k} + \lambda^j L^i_{jk} = -2\lambda^i \psi_k,$$

which is discussed in § 9.

If $\bar{\Gamma}^i_{jk}$ and $\bar{\Omega}^i_{jk}$ denote the symmetric and skew-symmetric parts of $\bar{L}^i_{jk}$, as in (3.3) and (5.1), we have

$$\bar{\Gamma}^i_{jk} = \Gamma^i_{jk} + \delta^i_j \psi_k + \delta^i_k \psi_j, \tag{12.3}$$

and

$$\bar{\Omega}^i_{jk} = \Omega^i_{jk} + \delta^i_j \psi_k - \delta^i_k \psi_j. \tag{12.4}$$

From the definition (§ 7) of the paths of a connected manifold it follows that the paths are the same for all connections related as in (12.1). This can be shown directly by means of (12.3). Conversely, if we apply to equations (7.6) the same reasoning as was applied to (7.5), we can show that expressions of the form (12.3) give the most general relation connecting the Γ's so that the equations (7.6) are unaltered. Hence we have:

Equations (12.3) *and an arbitrary choice of* Ω^i_{jk} *define the most general change in connection which preserves the paths.*

If L^i_{jkl} denote the components of the curvature tensor for the L's defined by (12.1), from (2.6) we have

$$L^i_{jkl} = L^i_{jkl} + 2\delta^i_j \left(\frac{\partial \psi_l}{\partial x^k} - \frac{\partial \psi_k}{\partial x^l} \right). \tag{12.5}$$

In like manner we have from (5.4) and (12.3)

(12.6) $\bar{B}^i_{jkl} = B^i_{jkl} + \delta^i_j(\psi_{lk} - \psi_{kl}) + \delta^i_l\,\psi_{jk} - \delta^i_k\,\psi_{jl}$,

where

(12.7) $$\psi_{jk} = \psi_{j,k} - \psi_j\,\psi_k,$$

$\psi_{j,k}$ being the covariant derivative of ψ_j with respect to the Γ's. From equations analogous to (5.3) and from (12.5) and (12.6) we have

(12.8) $\bar{\Omega}^i_{jkl} = \Omega^i_{jkl} + \delta^i_j(\psi_{lk} - \psi_{kl}) + \delta^i_k\,\psi_{jl} - \delta^i_l\,\psi_{jk}$.

From (12.5) by contraction we have

(12.9) $$\bar{L}_{jk} \equiv \bar{L}^i_{jki} = L_{jk} + 2\left(\frac{\partial\psi_j}{\partial x^k} - \frac{\partial\psi_k}{\partial x^j}\right)$$

and

(12.10) $$\bar{\Lambda}_{jk} \equiv \bar{L}^i_{ijk} = \Lambda_{jk} + 2n\left(\frac{\partial\psi_k}{\partial x^j} - \frac{\partial\psi_j}{\partial x^k}\right).$$

From this result and the theorem of § 5 we have:

The vector ψ_i can be chosen so that for the new linear connection $L^i_{ijk} = 0$.

From (12.5) we have:

When, and only when, ψ_k is a gradient, $\bar{L}^i_{jkl} = L^i_{jkl}$.

Contracting (12.6) for i and l and i and j, we have respectively

(12.11) $$\bar{B}_{jk} = B_{jk} + n\,\psi_{jk} - \psi_{kj}$$

and

(12.12) $$\bar{\beta}_{lk} = \beta_{lk} + \frac{n+1}{2}(\psi_{lk} - \psi_{kl}),$$

in consequence of (5.11). If in accordance with (5.10) we put

(12.13) $$\beta_{lk} = \frac{1}{2}\left(\frac{\partial\beta_l}{\partial x^k} - \frac{\partial\beta_k}{\partial x^l}\right),$$

we have from (12.12) that

(12.14) $$\bar{\beta}_l = \beta_l + (n+1)\,\psi_l + \sigma_l,$$

where σ_l is the gradient of an arbitrary function σ.

Again contracting (12.8) and using the notation of § 5, we have

$$\bar{\Omega}_{jk} = \Omega_{jk} + (2-n)\,\psi_{jk} - \psi_{kj} \tag{12.15}$$

and

$$\bar{\Phi}_{kl} = \Phi_{kl} + (n-1)\,(\psi_{lk} - \psi_{kl}). \tag{12.16}$$

From (12.4) and (5.12) we have

$$\bar{\Omega}_{k} = \Omega_{k} + (n-1)\,\psi_{k} \tag{12.17}$$

so that (12.16) is consistent with (5.14).

As an example of the second theorem of this section we consider the asymmetric connection which can be assigned to a Riemannian space so that the geodesics be the paths, that there be n independent vector-fields of parallel unit vectors and that the angle between two directions at a point and the parallel directions at any other point be equal.* In order that the first two conditions be satisfied we must have respectively

$$L^i_{jk} = \begin{Bmatrix} i \\ j\,k \end{Bmatrix} + \delta^i_j\,\psi_k + \delta^i_k\,\psi_j + \Omega^i_{jk}, \tag{12.18}$$

where the Christoffel symbols are given by (1.6), and $L^i_{jkl} = 0$ (§ 9), which in consequence of (5.3) and (6.2) are

$$R^i_{jkl} + \Omega^i_{jl,k} - \Omega^i_{jk,l} + \Omega^h_{jl}\,\Omega^i_{hk} - \Omega^h_{jk}\,\Omega^i_{hl} = 0, \tag{12.19}$$

where covariant differentiation is with respect to the g's and R^i_{jkl} are the components of the Riemannian curvature tensor. The third condition is $g_{ij|k} = 0$, which are reducible by means of (12.18) to

$$2\,g_{ij}\,\psi_k + g_{ik}\,\psi_j + g_{jk}\,\psi_i + \Omega_{jik} + \Omega_{ijk} = 0, \tag{12.20}$$

where

$$\Omega_{jik} = g_{jh}\,\Omega^h_{ik}.$$

Multiplying (12.20) by g^{ij} and summing for i and j, and by g^{ik} and summing for i and k, we find that

* Cf. *Cartan* and *Schouten*, 1926, 12.

(12.21) $$\psi_j = 0, \quad \Omega^i_{ij} = 0,$$

and hence from (12.20)

(12.22) $$\Omega_{jik} + \Omega_{ijk} = 0.$$

When we take the sum of the three equations obtained from (12.19) by permuting the subcripts cyclically and make use of known identities in the R's, we have

(12.23) $$\Omega^i_{jl,k} - \Omega^i_{jk,l} - \Omega^i_{kl,j} + \Omega^h_{jl}\,\Omega^i_{hk} - \Omega^h_{jk}\,\Omega^i_{hl} - \Omega^h_{kl}\,\Omega^i_{hj} = 0,$$

so that (12.19) may be written in the form

(12.24) $$R_{ijkl} + \Omega_{ikl,j} + \Omega^h_{kl}\,\Omega_{ihj} = 0.$$

From these equations because of (12.22) and well-known identities in R_{ijkl}* we obtain

$$R_{ijkl} = \frac{1}{3}(2\,\Omega_{hij}\,\Omega^h_{kl} + \Omega_{hik}\,\Omega^h_{jl} - \Omega_{hil}\,\Omega^h_{jk})$$

and hence from (12.24)

(12.25) $$\Omega_{ikl,j} = \frac{1}{3}(\Omega_{hij}\,\Omega^h_{kl} - \Omega_{hik}\,\Omega^h_{jl} + \Omega_{hil}\,\Omega^h_{jk}).$$

From (12.24) we have

$$R_{jk} = \Omega^h_{ik}\,\Omega^i_{hj}.$$

With the aid of (12.22) and (12.25) we obtain

(12.26) $$R_{jk,l} = 0.$$

A solution of these equations is furnished by the Einstein spaces, that is, spaces for which $R_{ij} = cg_{ij}$, where c is a constant. When this condition is not satisfied, it follows from (12.26) that the spaces are a sub-class of those considered by the author† (cf, § 29). For further considerations of the preceding case see the paper by Cartan and Schouten.

* 1926, 1, p. 21.
† Cf. 1923, 3.

13. Tensors independent of the choice of ψ_i. From (12.5), (12.9), (12.10), (12.4) and (12.17) it is seen that the following tensors are independent of the choice of the vector ψ_i in (12.1):

$$A^i_{jkl} = L^i_{jkl} + \delta^i_j L_{kl}, \tag{13.1}$$

$$L^i_{jkl} - \frac{1}{n}\delta^i_j A_{kl}, \tag{13.2}$$

$$L_{jk} + L_{kj}, \tag{13.3}$$

$$T^i_{jk} \equiv \Omega^i_{jk} + \frac{1}{n-1}(\delta^i_k \Omega_j - \delta^i_j \Omega_k). \tag{13.4}$$

From (12.15) and (12.16) we have

$$\psi_{jk} = -\frac{1}{n-1}\left[\Omega_{jk} - \Omega_{jk} + \frac{1}{n-1}(\Phi_{jk} - \Phi_{jk})\right],$$

$$\psi_{jk} - \psi_{kj} = \frac{1}{n-1}(\Phi_{kj} - \Phi_{kj}).$$

When these expressions are substituted in (12.8), we find that the tensor of components

$$\begin{aligned} T^i_{jkl} \equiv \Omega^i_{jkl} &+ \frac{1}{n-1}(\delta^i_k \Omega_{jl} - \delta^i_l \Omega_{jk} - \delta^i_j \Phi_{kl}) \\ &+ \frac{1}{(n-1)^2}(\delta^i_k \Phi_{jl} - \delta^i_l \Phi_{jk}) \end{aligned} \tag{13.5}$$

is independent of the choice of ψ_i.* From (13.5) we have by contraction

$$T^i_{jki} = 0, \qquad T^i_{ikl} = \frac{3-n}{(n-1)^2}\Phi_{kl} + \frac{1}{n-1}(\Omega_{kl} - \Omega_{lk}). \tag{13.6}$$

Other tensors independent of the choice of ψ_i are obtained in § 32. We close this section by establishing the following theorem:

A necessary and sufficient condition that a vector ψ_i can be chosen so that tensor L^i_{jkl} be zero is that the tensor A^i_{jkl} be zero.

* These results for (13.2), (13.4) and (13.5) are due to *J. M. Thomas*, 1926, 3, pp. 667, 668.

Evidently it is a necessary condition. Conversely, if the condition is satisfied, we have

$$L^i_{jkl} + \delta^i_j L_{kl} = 0. \tag{13.7}$$

Contracting for i and j, we have, using the notation of (12.10),

$$\Lambda_{kl} + n L_{kl} = 0.$$

From (5.3), (5.11), (5.14) and (12.13) we have

$$\Lambda_{kl} = \frac{\partial}{\partial x^k}(\beta_l + \Omega_l) - \frac{\partial}{\partial x^l}(\beta_k + \Omega_k). \tag{13.8}$$

Hence equations (13.7) become

$$L^i_{jkl} + \frac{1}{n}\delta^i_j\left[\frac{\partial}{\partial x^l}(\beta_k + \Omega_k) - \frac{\partial}{\partial x^k}(\beta_l + \Omega_l)\right] = 0.$$

Comparing these equations with (12.5) we see that $\overline{L}^i_{jkl} = 0$, if we take

$$\psi_k = -\frac{2}{n}(\beta_k + \Omega_k) + \frac{\partial \sigma}{\partial x^k},$$

where σ is any function of the x's.

14. Semi-symmetric connections. In § 12 it was shown that parallelism with respect to every curve in space cannot be the same for two symmetric connections. However, if for an asymmetric connection we have

$$\Omega^i_{jk} = \delta^i_j a_k - \delta^i_k a_j, \tag{14.1}$$

where a_j are the components of a vector, and we take $\psi_i = -a_i$, then $\overline{\Omega}^i_{jk} = 0$, as follows from (12.4). Conversely, in order that it be possible to choose ψ_i so that $\overline{\Omega}^i_{jk} = 0$, it is necessary that Ω^i_{jk} be of the form (14.1). Following Schouten* we say that the connection is *semi-symmetric* in this case. Hence we have:

* 1924, 1, p. 69.

A necessary and sufficient condition that parallelism be the same with respect to every curve for two connections one of which is symmetric is that the other be semi-symmetric.

From (12.4) it follows that, when a connection is semi-symmetric, the other connections with the same parallelism are semi-symmetric with the exception of a unique symmetric connection.

We establish the following theorem due to J. M. Thomas:*

A necessary and sufficient condition that an asymmetric connection be semi-symmetric is that there exist a coördinate system for each point of space in terms of which any vector at the point and that arising from it by a parallel displacement to any nearby point are proportional.

If such a coördinate system y^i exist and λ^i are the components of a vector at a point P, then at a nearby point the components are $\bar{\lambda}^i - \bar{L}^i_{jk}\bar{\lambda}^j dy^k$. The conditions of the theorem are given by

$$(\bar{L}^i_{jk}\delta^h_l - \bar{L}^h_{jk}\delta^i_l)\bar{\lambda}^j\bar{\lambda}^l dy^k = 0.$$

Proceeding with these equations in a manner analogous to that at the beginning of § 12, we obtain

$$\bar{L}^i_{jk} = \frac{\delta^i_j}{n}\bar{L}^h_{hk}. \tag{14.2}$$

From these equations we have

$$2\bar{\Omega}^i_{jk} = \frac{1}{n}(\delta^i_j\bar{L}^h_{hk} - \delta^i_k\bar{L}^h_{hj}).$$

Contracting for i and j, we have

$$2\bar{\Omega}_k = \frac{n-1}{n}\bar{L}^h_{hk} \tag{14.3}$$

and the preceding equations can be written as

$$\bar{\Omega}^i_{jk} = \frac{1}{n-1}(\delta^i_j\bar{\Omega}_k - \delta^i_k\bar{\Omega}_j). \tag{14.4}$$

Hence the connection must be semi-symmetric.

* 1926, 3, p. 670.

Conversely, if a connection is semi-symmetric and x^i are a general system of coördinates and P is the point of coördinates x^i, when we effect the transformation

$$(14.5)\quad x^i = x_0^i + y^i - \frac{1}{2}(\Gamma^i_{jk})_0\, y^j y^k + \frac{1}{n-1}(\Omega_j)_0\, y^j y^i,$$

we have at P

$$(14.6)\quad \begin{aligned} &\left(\frac{\partial \bar{x}^i}{\partial y}\right)_0 = \delta^i_{j\cdot}, \\ &\left(\frac{\partial^2 x^i}{\partial y^j\, \partial y^k}\right)_0 = -(\Gamma^i_{jk})_0 + \frac{1}{n-1}\left[\delta^i_j (\Omega_k)_0 + \delta^i_k (\Omega_j)_0\right] \end{aligned}$$

and from the first of these it follows that

$$(14.7)\qquad (\bar{\Omega}_j)_0 = (\Omega_j)_0.$$

Making use of equations of the form (2.1), we have,

$$(\bar{L}^i_{jk})_0 = (\Omega^i_{jk})_0 + \frac{1}{n-1}\left[\delta^i_j (\Omega_k)_0 + \delta^i_k (\Omega_j)_0\right].$$

Since equations (14.4) hold for any coördinate system, we have in consequence of (14.7),

$$(\bar{L}^i_{jk})_0 = \frac{2}{n-1}\, \delta^i_j\, (\bar{\Omega}_k)_0,$$

from which (14.3) follows by contraction. Hence in the coördinate system defined by (14.5), the conditions (14.2) are satisfied.

15. Transversals of parallelism of a given vector-field and associate vector-fields. If for a given vector-field λ^i the determinant $|\lambda^i_{\ |j}|$ is not zero, a necessary and sufficient condition that the determinant $|\bar{\lambda}^i_{\ |j}|$ for $\bar{\lambda}^i = \varphi \lambda^i$ be zero, that is, that the determinant

$$(15.1)\qquad \left|\varphi\, \lambda^i_{\ |j} + \lambda^i \frac{\partial \varphi}{\partial x^j}\right|$$

be zero, is that φ be a solution of the equation*

* Cf. *Kowalewski*, 1909, 2, p. 84; *Fine*, 1905, 1, p. 505.

$$(15.2) \qquad \begin{vmatrix} \varphi & -\frac{\partial \varphi}{\partial x^1} & \cdots & -\frac{\partial \varphi}{\partial x^n} \\ \lambda^1 & \lambda^1{}_{|1} & \cdots & \lambda^1{}_{|n} \\ \cdot & \cdot & \cdot & \cdot \\ \cdot & \cdot & \cdot & \cdot \\ \lambda^n & \lambda^n{}_{|1} & \cdots & \lambda^n{}_{|n} \end{vmatrix} = 0.$$

Moreover, the rank of (15.1) is $n-1$ for each solution. Hence in considering any vector-field we assume that the components are changed by a factor φ if necessary, so that $|\lambda^i{}_{|j}|$ is at most of rank $n-1$. We say that then the field is *normal* and that φ is the normalizing factor. This is a generalization of a unit, or a null, vector-field in a Riemannian space. For, in this case we have $\lambda_i \lambda^i{}_{,j} = 0$, and consequently $|\lambda^i{}_{|j}| = 0$.

If the rank of $|\lambda^i{}_{|j}|$ is $n-r$, there are r independent vector-fields $\mu^i_{(\alpha)}$ $(\alpha = 1, \cdots, r)$ which satisfy

$$(15.3) \qquad \mu^j \lambda^i{}_{|j} = 0$$

and the general solution of (15.3) is

$$(15.4) \qquad \mu^i = \psi^\alpha \mu^i_{(\alpha)} \qquad (\alpha = 1, \cdots, r),$$

where the ψ's are arbitrary functions of the x's.

When μ^i satisfy (15.3), the vectors λ^i are parallel with respect to each curve of the congruence defined by

$$\frac{dx^1}{\mu^1} = \cdots = \frac{dx^n}{\mu^n}$$

as follows from (9.1). Moreover, it follows that the vectors $\lambda^i \varphi$ are parallel, whatever be φ. Accordingly we say that each solution μ^i of (15.3) defines a congruence of *transversals of parallelism* of the field λ^i.*

When $|\lambda^i{}_{|j}|$ is of rank $n-r$, we say that the field λ^i is *general* or *special*, according as the rank of the matrix of the

* Transversals of parallelism for a surface in ordinary space were considered by *Bianchi*, 1923, 6, p. 806.

last n rows of (15.2) is $n-r+1$ or $n-r$. When the field is special, and also when it is general and $r>1$, equation (15.2) is satisfied by every function φ. When $r=1$ and the field is general, equation (15.2) reduces to

$$\mu^j \frac{\partial \varphi}{\partial x^j} = 0. \tag{15.5}$$

Suppose that the field is general and that φ is a solution of (15.5) when $r=1$, or any function whatever when $r>1$. The equations

$$\mu^j \left(\varphi \lambda^i_{\ |j} + \lambda^i \frac{\partial \varphi}{\partial x^j}\right) = 0 \tag{15.6}$$

are satisfied by all vectors μ^i defined by (15.4) for which the functions ψ^α satisfy the equation

$$\psi^\alpha \mu^j_{(\alpha)} \frac{\partial \varphi}{\partial x^j} = 0.$$

If there were a solution of (15.6) not expressible in the form (15.4), then from (15.6) we have equations of the form $\lambda^i = a^j \lambda^i_{\ |j}$, in which case the rank of the matrix of the last n rows of (15.2) is $n-r$. Hence when the field is general, all the solutions of (15.6) are expressible in the form (15.4), that is, on replacing λ^i by $\lambda^i \varphi$ no new congruences of transversals of parallelism are obtained.

When the field is special, the determinant (15.1) is of rank $n-r$ at most and there are at least r independent solutions of (15.6). Consequently if φ is such that not all of the equations

$$\mu^j_{(\alpha)} \frac{\partial \varphi}{\partial x^j} = 0 \qquad (\alpha = 1, \cdots, r)$$

are satisfied, there is another solution, say $\mu^i_{(\alpha+1)}$, of equations (15.6). Evidently it is such that $\mu^j_{(\alpha+1)} \frac{\partial \varphi}{\partial x^j} \neq 0$. If μ^i is any other solution of (15.6) not of the form (15.4), on

of solutions is determined by arbitrary values of λ^i for $x^1 = 0$, that is, by n arbitrary functions of $x^2, \cdots, x^n$. In particular, the n sets of solutions $\lambda^i_{(\alpha)}$, where α, for $\alpha = 1, \cdots, n$, determines the set and i the component, determined by the initial values $(\lambda^i_{(\alpha)})_0 = \delta^i_\alpha$, are independent, since the determinant $|\lambda^i_{(\alpha)}|$ is not identically zero. Moreover, from the form of (15.8) it follows that $\lambda^i = \varphi^\alpha \lambda^i_{(\alpha)}$ is also a solution, where the φ's are any functions of $x^2, \cdots, x^n$. Hence we have:

For any congruence μ^i there exist n independent vector-fields $\lambda^i_{(\alpha)}$ with respect to which the given congruence is the congruence of transversals of parallelism; moreover, the field

$$\lambda^i = \varphi^\alpha \lambda^i_{(\alpha)} \qquad (\alpha = 1, \cdots, n)$$

possesses the same property, when the φ's are any solutions of the equation

$$\mu^i \frac{\partial \varphi}{\partial x^i} = 0,$$

the coördinates x^i being any whatever.

When $|\lambda^i{}_{|j}|$ is of rank $n - r$, the equations

$$\nu_i \lambda^i{}_{|j} = 0 \tag{15.9}$$

are satisfied by r independent covariant vector-fields $\nu^{(\alpha)}_i$ $(\alpha = 1, \cdots, r)$ and the general solution is

$$\nu_i = \psi_\alpha \nu^{(\alpha)}_i, \tag{15.10}$$

where the ψ's are arbitrary functions of the x's. We say that each such field is *associate* to the given field λ^i.* If the given field is general, there are $r - 1$ fields of independent vectors given by (15.10) for which $\nu_i \lambda^i = 0$, and these fields are associate to the field $\lambda^i \varphi$ for every φ satisfying (15.2). If, however, the field is special each of the fields (15.10) is associate to $\lambda^i \varphi$, whatever be φ.

* By normalizing the field we have that the rank is at most $n - 1$ and consequently there is at least one associate covariant field.

eliminating λ^i from (15.6) and from the similar equati[on] when μ^j is replaced by $\mu^j_{(\alpha+1)}$, we have

$$(15.7)\quad \left(\mu^j_{(\alpha+1)}\,\mu^k\,\frac{\partial \log \varphi}{\partial x^k} - \mu^j\,\mu^k_{(\alpha+1)}\,\frac{\partial \log \varphi}{\partial x^k}\right)\lambda^i_{\,|j} = 0,$$

and consequently μ^i is expressible linearly in terms of $\mu^i_{(\beta)}$ $(\beta = 1, \cdots, r+1)$. Hence for the given function φ all solutions of (15.6) are expressible linearly in terms of these $r+1$ vectors. For another function, say φ_1, there is at most one field $\mu^i_{(\alpha+2)}$ other than $\mu^i_{(\alpha)}$ $(\alpha = 1, \cdots, r)$. But in this case we have the equations obtained from (15.7) on replacing φ in the first term of the left-hand member by φ_1 and μ^j throughout by $\mu^j_{(\alpha+2)}$. Consequently the change of the function φ does not yield new congruences of transversals of parallelism.

Gathering these results together, we have:

When a vector-field λ^i is normal and the rank of $|\lambda^i_{\,|j}|$ is $n-r$, there are r independent congruences of transversals of parallelism, unless the rank of the matrix of the last n rows of (15.2) *is $n-r$; in the latter case there are $r+1$ independent congruences of transversals; moreover, in either case any linear combination of the vectors defining congruences of transversals also defines such a congruence.*

When $\lambda^i_{\,|j} = \lambda^i \sigma_j$, where σ_j is any vector, the vectors λ^i are parallel with respect to any curve in the space (cf. §§ 9, 10).

We consider the converse problem: Given a vector-field μ^i to determine the vector-fields λ^i for which the former is a congruence of transversals of parallelism. We assume that the coördinate system x^i is that for which $\mu^\sigma = 0$ $(\sigma = 2, \cdots, n)$.* In this coördinate system the equations (15.3) for the determination of the λ's are

$$(15.8)\qquad \frac{\partial \lambda^i}{\partial x^1} + \lambda^j L^i_{j1} = 0.$$

Any set of functions λ^i satisfying these equations are the components in the x's of a vector-field with respect to which the congruence μ^i is the congruence of transversals. A set

* 1926, 1, p. 5.

In like manner the components μ_i of a covariant field can be chosen so that the rank of $|\mu_{i|j}|$ is $n-r$ $(r \geqq 1)$. Any solution λ^i of

$$\lambda^i \mu_{i|j} = 0$$

gives the components of a field of contravariant vectors *associate* to the given field.*

16. Associate directions. Consider a field of non-parallel contravariant vectors of components λ^i and a curve C at points of which the coördinates x^i are expressed in terms of a parameter t. A family of contravariant vectors of components μ^i is defined at points of C by the equations

$$\frac{dx^j}{dt} \lambda^i_{\ |j} = \mu^i. \tag{16.1}$$

If $\mu^i = f(t)\lambda^i$, the vectors are parallel with respect to C. When this condition is not satisfied, we say that μ^i are the components of the *associate direction* of λ^i with respect to C.

If λ^i are replaced in (16.1) by $\lambda^i \varphi$, where φ is any function of the x's, and $\overline{\mu}^i$ are the components of the associate directions of the latter vectors, we have

$$\overline{\mu}^i = \varphi \mu^i + \lambda^i \frac{d\varphi}{dt}. \tag{16.2}$$

In this way we get a pencil of associate directions, determined by the given vector and any one of the associate directions. Conversely it is possible to choose a function φ such that the associate direction of $\varphi \lambda^i$ is a given one of the pencil other than the direction λ^i.

When the given vector-field has been normalized (§ 15), if necessary, and ν_i are the components of an associate covariant vector, we have $\nu_i \mu^i = 0$. Hence we have:

For a field of non-parallel contravariant vectors the associate directions with respect to a curve are pseudo-örthogonal to the associate covariant vectors of the field.

* *Eisenhart*, 1926, 14.

In particular, if C is not a path of the space and λ^i are the components of the tangent to C, that is, $\lambda^i = \frac{dx^i}{dt}$, equations (16.1) become

$$\frac{d^2x^i}{dt^2} + \Gamma^i_{jk}\frac{dx^j}{dt}\frac{dx^k}{dt} = \mu^i. \tag{16.3}$$

If we change the parameter t, we get a pencil of associate directions as in (16.2). We note that associate directions of a curve are independent of the tensor Ω^i_{jk}. The associate directions of the tangent are evidently a generalization of the pencil determined by the tangent and first curvature normal of a curve in a Riemannian space (cf. § 24).*

In a similar manner, if λ_i are the components of any field of non-parallel covariant vectors, the equations

$$\lambda_{i|j}\frac{dx^j}{dt} = \mu_i$$

define the associate covariant vector μ_i of λ_i with respect to the curve, unless the vectors λ_i are parallel with respect to it, that is, unless $\mu_i = \lambda_i f(t)$. When λ_i is replaced by $\varphi\lambda_i$, where φ is an arbitrary function of the x's, we get a pencil of associate covariant vectors determined by the given vector and any one of them. Moreover, we have:

For a field of non-parallel covariant vectors the associate covariant vectors with respect to a curve are pseudo-örthogonal to the associate contravariant vectors of the field.

17. Determination of a tensor by an ennuple of vectors and invariants. Let $\lambda^i_{(\alpha)}$ denote the components† of n linearly independent vectors in a coördinate system x^i. Then the determinant

$$\lambda = |\lambda^i_{(\alpha)}| \tag{17.1}$$

* 1926, 1, pp. 60, 72.

† As formerly the index with parentheses indicates the vector and the one without parentheses the component. This convention will be followed hereafter, and unless stated otherwise the indices take the value $1, \cdots, n$; moreover, the summation convention is used for both sets of indices.

is not identically zero. We denote by $\lambda_i^{(\alpha)}$ the n^2 functions defined by the equations

$$(17.2) \qquad \lambda_i^{(\alpha)} \lambda_{(\beta)}^i = \delta_\beta^\alpha;$$

as thus defined $\lambda_i^{(\alpha)}$ is the cofactor $\lambda_{(\alpha)}^i$ in λ divided by λ. In any other coördinate system x'^i the functions $\lambda'^{(\alpha)}_i$ defined by $\lambda'^{(\alpha)}_i \lambda'^i_{(\beta)} = \delta_\beta^\alpha$ are such that

$$\lambda'^{(\alpha)}_i = \lambda_j^{(\alpha)} \frac{\partial x^j}{\partial x'^i}.$$

Consequently $\lambda_i^{(\alpha)}$ are the components of n independent covariant vectors. Furthermore, it follows from (17.2) that

$$(17.3) \qquad \lambda_i^{(\alpha)} \lambda_{(\alpha)}^j = \delta_i^j.$$

If we had started with the independent covariant vectors $\lambda_i^{(\alpha)}$, then equations (17.2) serve to define n independent contravariant vectors. Owing to the reciprocal character of the relations (17.2), we say that either set is *conjugate* to the other, and that the two sets constitute an *ennuple.* It is evident that an orthogonal ennuple of contravariant vectors in a Riemannian space* and the associate covariant vectors form an ennuple in the above sense.

If $a_{j_1 \cdots j_s}^{i_1 \cdots i_r}$ are the components of a tensor, then the quantities

$$(17.4) \qquad c_{\alpha_1 \cdots \alpha_s}^{\beta_1 \cdots \beta_r} = a_{j_1 \cdots j_s}^{i_1 \cdots i_r} \lambda_{i_1}^{(\beta_1)} \cdots \lambda_{i_r}^{(\beta_r)} \lambda_{(\alpha_1)}^{j_1} \cdots \lambda_{(\alpha_1)}^{j_s}$$

are invariants. If these expressions are substituted in the right-hand members of the equations

$$(17.5) \qquad a_{j_1 \cdots j_s}^{i_1 \cdots i_r} = c_{\alpha_1 \cdots \alpha_s}^{\beta_1 \cdots \beta_r} \lambda_{(\beta_1)}^{i_1} \cdots \lambda_{(\beta_r)}^{i_r} \lambda_{j_1}^{(\alpha_1)} \cdots \lambda_{j_s}^{(\alpha_s)},$$

these equations are identically satisfied because of (17.3). Hence we have:

* 1926, 1, pp. 14, 40, 96.

*The components of any tensor are expressible in terms of invariants and the components of an ennuple.**

In particular, we can express Ω^k_{ij} in the form

$$\Omega^k_{ij} = \omega^{\alpha}_{\gamma\beta}\,\lambda^k_{(\alpha)}\,\lambda^{(\beta)}_i\,\lambda^{(\gamma)}_j, \tag{17.6}$$

where $\omega^{\alpha}_{\beta\gamma}$ are skew-symmetric in the subscripts.

We shall apply the preceding results to show that, if a_{ijk} is a tensor such that $a_{ijk}\,\lambda^i_{(\varrho)}\,\lambda^j_{(\sigma)}\,\lambda^k_{(\tau)} = 0$ for ϱ, σ and τ not equal to n, then

$$a_{ijk} = \lambda^{(n)}_i\,a_{jk} + \lambda^{(n)}_j\,b_{ki} + \lambda^{(n)}_k\,c_{ij}, \tag{17.7}$$

where a_{ij}, b_{ij}, c_{ij} are tensors. In fact, if we write a_{ijk} in the form (17.5), that is

$$a_{ijk} = c_{\alpha\beta\gamma}\,\lambda^{(\alpha)}_i\,\lambda^{(\beta)}_j\,\lambda^{(\gamma)}_k,$$

we have that $c_{\varrho\sigma\tau} = 0$. Hence (17.7) follows, where $a_{jk} = c_{n\alpha\beta}\,\lambda^{(\alpha)}_j\,\lambda^{(\beta)}_k$ and so on.

Any other ennuple $\overline{\lambda}^i_{(\sigma)}$, $\overline{\lambda}^{(\sigma)}_i$ is given by

$$\overline{\lambda}^i_{(\sigma)} = a^{\alpha}_{\sigma}\,\lambda^i_{(\alpha)}, \quad \overline{\lambda}^{(\sigma)}_i = A^{\sigma}_{\alpha}\,\lambda^{(\alpha)}_i, \tag{17.8}$$

where the determinant $|a^{\alpha}_{\sigma}|$ is not identically zero, and the a's and A's are invariants in the relations

$$a^{\alpha}_{\sigma}\,A^{\tau}_{\alpha} = \delta^{\tau}_{\sigma}, \tag{17.9}$$

as follows from (17.2). If $\overline{c}^{\sigma_1\cdots\sigma_r}_{\tau_1\cdots\tau_s}$ are the invariants for the tensor $a^{i_1\cdots i_r}_{j_1\cdots j_s}$ with respect to this ennuple, we have

$$\overline{c}^{\sigma_1\cdots\sigma_r}_{\tau_1\cdots\tau_s} = c^{\alpha_1\cdots\alpha_r}_{\beta_1\cdots\beta_s}\,A^{\sigma_1}_{\alpha_1}\cdots A^{\sigma_r}_{\alpha_r}\,a^{\beta_1}_{\tau_1}\cdots a^{\beta_s}_{\tau_s}. \tag{17.10}$$

When for a given coördinate system we take

$$\lambda^i_{(\alpha)} = \delta^i_{\alpha}, \tag{17.11}$$

* Cf. 1926, 1, p. 97.

then

(17.12) $$\lambda_i^{(\alpha)} = \delta_i^\alpha ,$$

as follows from (17.2) or (17.3). For this particular ennuple we have from (17.5)

(17.13) $$a^{i_1\cdots i_r}_{j_1\cdots j_s} = c^{i_1\cdots i_r}_{j_1\cdots j_s},$$

that is, any component of the tensor in this coördinate system is equal to the invariant with the same indices as the component. We call the ennuple (17.11) and (17.12) the *fundamental ennuple* of the given coördinate system.

18. The invariants $\gamma_{\mu}{}^{\nu}{}_{\sigma}$ of an ennuple. For a given ennuple the invariants $\gamma_{\mu}{}^{\nu}{}_{\sigma}$ defined by

(18.1) $$\gamma_{\mu}{}^{\nu}{}_{\sigma} = \lambda^i_{(\mu)|j}\,\lambda_i^{(\nu)}\,\lambda^j_{(\sigma)}$$

are a generalization of the coefficients of rotation of an orthogonal ennuple in a Riemannian space, as defined by Ricci and Levi-Civita.*

From (18.1) we have because of (17.3)

(18.2) $$\lambda^i_{(\mu)|j} = \gamma_{\mu}{}^{\nu}{}_{\sigma}\,\lambda^i_{(\nu)}\,\lambda_j^{(\sigma)}.$$

If equations (18.2) be multiplied by $\lambda_h^{(\mu)}$ and summed for μ, the resulting equations are reducible by means of (2.2) to

(18.3) $$L^i_{hj} = -\lambda_h^{(\alpha)}\frac{\partial \lambda^i_{(\alpha)}}{\partial x^j} + \gamma_{\mu}{}^{\nu}{}_{\sigma}\,\lambda_h^{(\mu)}\,\lambda^i_{(\nu)}\,\lambda_j^{(\sigma)}.$$

Conversely, if we have any ennuple and a set of invariants $\gamma_{\mu}{}^{\nu}{}_{\sigma}$ and we define functions L^i_{hj} by (18.3) and L'^i_{hj} by corresponding equations for any other coördinate system x'^i, we find that equations (2.1) are satisfied. Hence we have:

An ennuple of vectors and any set of invariants $\gamma_{\mu}{}^{\nu}{}_{\sigma}$ determine a connection; and any asymmetric connection is so determined.†

* 1901, 1, p. 148; cf. 1926, 1, p. 97.

† Cf. *Levy*, 1927, 1, p. 307.

When in particular, we take $\gamma_{\mu}{}^{\nu}{}_{\sigma} = 0$, we have from (18.3)

$$L^{i}_{hj} = -\lambda^{(\alpha)}_{h} \frac{\partial \lambda^{i}_{(\alpha)}}{\partial x^{j}} \tag{18.4}$$

and from (18.2)

$$\lambda^{i}_{(\alpha)|j} = 0.* \tag{18.5}$$

Consequently the n fields of vectors $\lambda^{i}_{(\alpha)}$ are parallel fields and hence $L^{i}_{jkl} = 0$ (§ 9). Conversely, if the latter condition is satisfied, we can choose n linearly independent vector-fields satisfying (18.5) and consequently we have (18.4). Hence we have:

A necessary and sufficient condition that the curvature tensor L^{i}_{jkl} of a manifold with asymmetric connection be zero is that the coefficients of the connection be expressible in the form (18.4) *in terms of an ennuple.*

From the form of equations (2.1) we have:

If L^{i}_{jk} are the coefficients of a connection, so also are $L^{i}_{jk} + a^{i}_{jk}$, where a^{i}_{jk} are the components of an arbitrary tensor.

As a consequence of this theorem we have that for any ennuple the quantities

$$\bar{L}^{i}_{hj} = -\lambda^{(\alpha)}_{j} \frac{\partial \lambda^{i}_{(\alpha)}}{\partial x^{h}} \tag{18.6}$$

are the coefficients of an asymmetric connection. For from (18.4) and (3.3) we have

$$\bar{L}^{i}_{hj} = L^{i}_{jh} = L^{i}_{hj} + 2\,\Omega^{i}_{jh}. \tag{18.7}$$

For the connection defined by (18.6) we have from (18.3)

$$\gamma_{\mu}{}^{\nu}{}_{\sigma} = \lambda^{(\nu)}_{i} \left(\lambda^{j}_{(\sigma)} \frac{\partial \lambda^{i}_{(\mu)}}{\partial x^{j}} - \lambda^{j}_{(\mu)} \frac{\partial \lambda^{i}_{(\sigma)}}{\partial x^{j}} \right), \tag{18.8}$$

from which it follows that $\gamma_{\mu}{}^{\nu}{}_{\sigma}$ is skew-symmetric in μ and σ.

Equations (18.7) show that for the connections (18.4) and (18.6) to be the same, it is necessary that L^{i}_{hj} be symmetric in h and j. In this case equations (18.5) become

* Cf. *Weitzenböck*, 1923, 2, p. 319.

$$\lambda^i_{(\alpha),j} = 0, \tag{18.9}$$

where the covariant differentiation is with respect to the Γ's. Then from (9.6) and (5.3) we have $B^i_{jkl} = 0$. Conversely, when these conditions are satisfied, equations (18.9) admit n linearly independent fields of vectors parallel with respect to the Γ's.

From (18.3) it follows that a necessary and sufficient condition that $\gamma_{\mu}{}^{\nu}{}_{\sigma}$ be skew-symmetric in the indices μ and σ is that

$$L^i_{hj} + L^i_{jh} = -\left(\lambda^{(\alpha)}_h \frac{\partial \lambda^i_{(\alpha)}}{\partial x^j} + \lambda^{(\alpha)}_j \frac{\partial \lambda^i_{(\alpha)}}{\partial x^h}\right).$$

This is the symmetric part of either of the coefficients (18.4) or (18.6) and consequently satisfies (5.6). In consequence of this result and the above theorem we have:

A necessary and sufficient condition that the invariants $\gamma_{\mu}{}^{\nu}{}_{\sigma}$ be skew-symmetric in the indices μ and σ is that

$$L^i_{hj} = -\frac{1}{2}\left(\lambda^{(\alpha)}_h \frac{\partial \lambda^i_{(\alpha)}}{\partial x^j} + \lambda^{(\alpha)}_j \frac{\partial \lambda^i_{(\alpha)}}{\partial x^h}\right) + \Omega^i_{hj}, \tag{18.10}$$

where Ω^i_{hj} is an arbitrary tensor skew-symmetric in h and j.

If we denote by $\lambda^i_{(\alpha)|\bar{k}}$ the covariant derivative of $\lambda^i_{(\alpha)}$ for the connection $\bar{L}^i_{jk}$ defined by (12.1), we have

$$\lambda^i_{(\alpha)|\bar{k}} - \lambda^i_{(\alpha)|k} = 2\psi_k \lambda^i_{(\alpha)},$$

from which it follows that

$$\lambda^{(\alpha)}_i (\lambda^i_{(\alpha)|\bar{k}} - \lambda^i_{(\alpha)|k}) = 2n\psi_k.$$

Consequently the mixed tensor

$$\lambda^i_{(\alpha)|k} - \frac{1}{n}\lambda^i_{(\alpha)} \lambda^{(\beta)}_j \lambda^j_{(\beta)|k} \tag{18.11}$$

is independent of the choice of the vector ψ_i. The same is true of the tensor

$$\lambda^{(\alpha)}_{i|k} - \frac{1}{n}\lambda^{(\alpha)}_{i}\lambda^{j}_{(\beta)}\lambda^{(\beta)}_{j|k}.$$

If equations (18.11) be multiplied by $\lambda^{(\nu)}_i \lambda^k_{(\sigma)}$, and i and k be summed, we find that the invariants

$$\gamma_{\alpha}{}^{\nu}{}_{\sigma} - \frac{1}{n}\delta^{\nu}_{\alpha}\gamma_{\beta}{}^{\beta}{}_{\sigma} \tag{18.12}$$

are independent of the choice of the vector ψ_i in (12.1). Conversely, if we have two asymmetric connections L^i_{jk} and $\bar{L}^i_{jk}$ for which the invariants (18.12) are equal for a given ennuple, it follows from (18.3) that

$$\bar{L}^i_{jk} - L^i_{jk} = \frac{1}{n}\delta^{\,i}_{j}(\bar{\gamma}_{\beta}{}^{\beta}{}_{\sigma} - \gamma_{\beta}{}^{\beta}{}_{\sigma})\lambda^{(\sigma)}_k,$$

which evidently are of the form (12.1). Hence we have:

A necessary and sufficient condition that parallelism be the same for two different asymmetric connections is that the corresponding invariants (18.12) *for a given ennuple be equal for these connections.*

19. Geometric properties expressed in terms of the invariants $\gamma_{\mu}{}^{\nu}{}_{\sigma}$. In order that the vector-field $\lambda^i_{(\alpha)}$ of an ennuple at points of each curve of a congruence $\lambda^i_{(\beta)}$ be parallel with respect to the curve, it is necessary and sufficient that

$$\lambda^j_{(\beta)}\lambda^i_{(\alpha)|j} = \varrho\lambda^i_{(\alpha)}.$$

By means of (18.2) these equations are equivalent to

$$(\gamma_{\alpha}{}^{\nu}{}_{\beta} - \varrho\delta^{\nu}_{\alpha})\lambda^i_{(\nu)} = 0.$$

Hence we have:

A necessary and sufficient condition that the vector-field $\lambda^i_{(\alpha)}$ of an ennuple be parallel with respect to the curves of a congruence $\lambda^i_{(\beta)}$ is that

$$\gamma_{\alpha}{}^{\nu}{}_{\beta} = 0 \quad (\nu = 1, \cdots, n;\ \nu \neq \alpha). \tag{19.1}$$

As a corollary we have:

A necessary and sufficient condition that the curves of the congruence $\lambda^i_{(\alpha)}$ of an ennuple be paths is that

$$\gamma_{\alpha\ \alpha}^{\ \nu} = 0^* \quad (\nu = 1, \cdots, n;\ \nu \neq \alpha). \tag{19.2}$$

If we use the notation

$$\frac{\partial f}{\partial t^\alpha} \equiv \lambda^i_{(\alpha)} \frac{\partial f}{\partial x^i}, \tag{19.3}$$

then

$$\begin{aligned}
&\left(\frac{\partial}{\partial t^\alpha}\frac{\partial}{\partial t^\beta} - \frac{\partial}{\partial t^\beta}\frac{\partial}{\partial t^\alpha}\right) f \\
&= \frac{\partial f}{\partial x^j}\left(\lambda^i_{(\alpha)}\lambda^j_{(\beta)/i} - \lambda^i_{(\beta)}\lambda^j_{(\alpha)/i} + 2\lambda^i_{(\alpha)}\lambda^k_{(\beta)}\Omega^j_{ik}\right) \\
&= \left(\gamma_{\beta\ \alpha}^{\ \nu} - \gamma_{\alpha\ \beta}^{\ \nu} + 2\omega_{\alpha\ \beta}^{\ \nu}\right)\frac{\partial f}{\partial t^\nu},
\end{aligned} \tag{19.4}$$

in consequence of (18.2) and (17.6). These equations are generalizations of equations due to Ricci and Levi-Civita† in Riemannian geometry. As an application of these equations we seek necessary and sufficient conditions that p of the congruences of an ennuple, say $\lambda^i_{(\sigma)}$ $(\sigma = 1, \cdots, p)$, generate a system of ∞^{n-p} varieties V_p. In this case the equations

$$\lambda^i_{(\sigma)} \frac{\partial f}{\partial x^i} = 0 \qquad (\sigma = 1, \cdots, p) \tag{19.5}$$

must form a complete system. From (19.4) we have:

A necessary and sufficient condition that the congruences $\lambda^i_{(\sigma)}$ for $\sigma = 1, \cdots, p$ generate a system of ∞^{n-p} varieties V_p is that

$$\gamma_{\beta\ \alpha}^{\ \nu} - \gamma_{\alpha\ \beta}^{\ \nu} + 2\omega^{\nu}_{\alpha\beta} = 0 \tag{19.6}$$
$$(\alpha, \beta = 1, \cdots, p;\ \nu = p+1, \cdots, n).$$

As a corollary we have:

A necessary and sufficient condition that there exist a coördinate system such that the curves of the congruences of

* Cf. 1926, 1, p. 100: also, *Levy*, 1927, 1, p. 308.

† 1901, 1, p. 150: cf. also 1926, 1, p. 99.

an ennuple be coördinate lines is that equations (19.6) *hold for all distinct values of* α, β *and* ν.*

We say that a congruence λ_i is *pseudonormal* to a family of hypersurfaces $f(x^1, \cdots, x^n) = \text{const.}$, if

$$\lambda_i = \mu \frac{\partial f}{\partial x^i}.$$

From the preceding results we have:

A necessary and sufficient condition that a congruence $\lambda_i^{(n)}$ *of an ennuple be pseudonormal to a family of hypersurfaces is that*

$$\gamma_{\alpha}{}^{n}{}_{\beta} - \gamma_{\beta}{}^{n}{}_{\alpha} + 2\omega^{n}_{\alpha\beta} = 0 \quad (\alpha, \beta = 1, \cdots, n-1). \dagger \tag{19.7}$$

* These two theorems for the case of a symmetric connection, in which case $\omega^{\nu}_{\alpha\beta} = 0$, are due to *Levy*, 1927, 1, p. 308.

† Cf. 1926, 1, p. 115.

CHAPTER II

SYMMETRIC CONNECTIONS

20. Geodesic coördinates. When a coördinate system can be chosen for which the coefficients of the connection vanish at a given point $P(x_0^i)$, the vector at any nearby point $P'(x_0^i + dx^i)$ parallel to a given contravariant vector at P has the same components as at P to within terms of the second and higher orders, as follows from (7.1). If in equations (2.1) we put $L^i_{jk} = 0$, we see that a necessary condition is that the coefficients in any other coördinate system be symmetric at P.

In order to show that this condition is also sufficient, we imagine that the space is referred to a general coördinate system x^i and we consider the transformation of coördinates defined by

$$x^i = x_0^i + \delta^i_j x'^j - \frac{1}{2}(\Gamma^i_{kl})_0 x'^k x'^l + \psi^i, \tag{20.1}$$

where ψ^i are any functions of the x''s such that they and their first and second derivatives are zero when the x''s are zero.* From (20.1) we have at P

$$\left(\frac{\partial x^i}{\partial x'^j}\right)_0 = \delta^i_j. \qquad \left(\frac{\partial^2 x^i}{\partial x'^j \partial x'^k}\right)_0 = -(\Gamma^i_{jk})_0. \tag{20.2}$$

From these expressions and equations (5.6) we have at P

$$(\Gamma'^i_{jk})_0 = 0. \tag{20.3}$$

Consequently any coördinate system defined by (20.1) possesses the desired property. Hence we have:

When, and only when, at a point the coefficients of a connection are symmetric in the subscripts, coördinate systems can

* Cf. 1926, 1, p. 56.

be chosen, with the point as origin, such that the coefficients are zero at the point.

Weyl* calls a connection affine, when at every point a coördinate system exists for which the components of a vector in this coördinate system remain unaltered by an infinitesimal displacement, to within terms of the second and higher orders, but we use the term affine for asymmetric connections as well (cf. § 7).

Any coördinate system for which (20.3) is satisfied has been called *geodesic* by Weyl. From the foregoing results it follows that if the coördinates x^i are geodesic for a point P as origin, other geodesic coördinate systems with the same origin are defined by

$$x^i = \delta^i_j x'^j + \psi^i. \tag{20.4}$$

where the ψ's are of the character appearing in (20.1).

It is evident that at the origin of a geodesic coördinate system first covariant derivatives reduce to ordinary derivatives. Consequently the use of such a system frequently makes for considerable simplification in any problem involving first covariant derivatives. Moreover, when the results of such an investigation are stated in tensor form, their generality is not conditioned by the use of the particular coördinate system.

Symmetric connections are characterized by another property. Consider a point $P(x^i)$ and two infinitesimal vectors $d_1 x^i$ and $d_2 x^i$ at P, and denote by P_1 and P_2 the points of coördinates $x^i + d_1 x^i$ and $x^i + d_2 x^i$ respectively. When the vector $d_1 x^i$ undergoes a general parallel displacement to P_2, its components at P_2 are $d_1 x^i + d_2 d_1 x^i + L^i_{jk} d_1 x^j d_2 x^k$, and the coördinates of the point P_{21} at the extremity of the vector are

$$x^i + d_2 x^i + d_1 x^i + d_2 d_1 x^i + L^i_{jk} d_1 x^j d_2 x^k.$$

In like manner when the vector $d_2 x^i$ undergoes a parallel displacement to P_1, the coördinates of the point P_{12} at the extremity of the vector are

$$x^i + d_1 x^i + d_2 x^i + d_1 d_2 x^i + L^i_{jk} d_2 x^j d_1 x^k.$$

* 1921, 1, p. 112.

Hence a necessary and sufficient condition that P_{12} and P_{21} coincide is that L^i_{jk} be symmetric in j and k, that is, that the connection be symmetric.*

21. The curvature tensor and other fundamental tensors. In § 5 it was seen that the quantities

$$B^i_{jkl} = \frac{\partial \Gamma^i_{jl}}{\partial x^k} - \frac{\partial \Gamma^i_{jk}}{\partial x^l} + \Gamma^h_{jl}\Gamma^i_{hk} - \Gamma^h_{jk}\Gamma^i_{hl} \tag{21.1}$$

are the components of a tensor. This tensor arises when we express the conditions of integrability of equations (5.6). In fact, these conditions assume the form

$$\frac{\partial x'^p}{\partial x^j}\frac{\partial x'^q}{\partial x^k}\frac{\partial x'^r}{\partial x^l} B'^s_{pqr} = \frac{\partial x'^s}{\partial x^i} B^i_{jkl}, \tag{21.2}$$

from which equations the tensor character of B^i_{jkl} is apparent. This tensor is a generalization of the Riemannian curvature tensor of a Riemannian space and we call it the *curvature tensor* of the space with symmetric connection.

From (21.1) it follows that B^i_{jkl} is skew-symmetric in k and l, that is,

$$B^i_{jkl} + B^i_{jlk} = 0. \tag{21.3}$$

Also the components satisfy the identities

$$B^i_{jkl} + B^i_{klj} + B^i_{ljk} = 0. \tag{21.4}$$

This result is readily proved by choosing a geodesic coördinate system at a point P. In this case at P all of the Γ's are zero and (21.4) can be shown to hold at P in this coördinate system. Since this is a tensor equation it holds at P in any coördinate system. Moreover, as P is any point, it holds throughout the space.

In like manner at a point P in a geodesic coördinate system with P origin, we have

* Cf. *Weyl*, 1921, 1, p. 107; *Levi-Civita*, 1925, 5, p. 135.

$$B^i_{jkl,m} = \frac{\partial^2 \Gamma^i_{jl}}{\partial x^k \partial x^m} - \frac{\partial^2 \Gamma^i_{jk}}{\partial x^l \partial x^m},$$

as follows from (21.1) and (6.1). Consequently

$$B^i_{jkl,m} + B^i_{jlm,k} + B^i_{jmk,l} = 0. \tag{21.5}$$

Since these are tensor equations, they hold throughout space in any coördinate system. They are evidently a generalization of the identities of Bianchi for a Riemannian space, and are called the *identities of Bianchi for a symmetric connection.**

In a similar manner the following identities due to Veblen† can be established:

$$B^i_{jkl,m} + B^i_{ljm,k} + B^i_{mlk,j} + B^i_{kmj,l} = 0.‡$$

22. Equations of the paths. In § 12 it was shown that parallelism throughout a space with symmetric connection is uniquely defined, that is, that it is not possible to have two symmetric connections with respect to which parallel directions along every curve in the space are the same for both connections; thus each symmetric connection is a unique affine connection. However, as a corollary of the third theorem of § 12 we have:

The paths are the same for two symmetric connections whose coefficients are in the relations

$$\bar{\Gamma}^i_{jk} = \Gamma^i_{jk} + \delta^i_j\, \psi_k + \delta^i_k\, \psi_j, \tag{22.1}$$

where ψ_j is an arbitrary covariant vector.§

The paths are a generalization of the straight lines of euclidean space. Accordingly the properties of the space

* Cf. *Veblen*, 1922, 5, p. 197; *Schouten*, 1923, 7.

† L. c., p. 197.

‡ It is evident that the results of these two sections apply not only to the case of symmetric connections, but that they apply also to the symmetric parts of any asymmetric connection.

§ *Weyl*, 1921, 2, p. 100; cf. also *Eisenhart*, 1922, 2 and *Veblen*, 1922, 3.

which depend upon the paths and not upon a particular affine connection of the set (22.1) constitute a *projective geometry of paths,* whereas those depending upon a particular affine connection constitute an *affine geometry of paths.* In this chapter we consider the latter and postpone to the next chapter a study of the former.

If we have a particular path, that is, an integral curve of equations (7.6), then

$$\text{(22.2)} \qquad \frac{d^2 x^i}{dt^2} + \Gamma^i_{jk} \frac{dx^j}{dt} \frac{dx^k}{dt} = \varphi \frac{dx^i}{dt}.$$

where φ is a determinate function of t. If we define a parameter s by

$$\text{(22.3)} \qquad \frac{ds}{dt} = c e^{\int \varphi \, dt},$$

where c is an arbitrary constant, equations (22.2) become

$$\text{(22.4)} \qquad \frac{d^2 x^i}{ds^2} + \Gamma^i_{jk} \frac{dx^j}{ds} \frac{dx^k}{ds} = 0.$$

Thus the parameter s for a path, which we call an *affine parameter*, is the analogue of the arc s of a geodesic in a Riemannian space.* It is evident from (22.3) or (22.4) that, if s is any affine parameter, the most general one is given by $as + b$ where a and b are arbitrary constants. Furthermore, by means of equations (5.6) we can establish the theorem (cf. § 38):

When the coördinates x^i undergo a general transformation, an affine parameter is not altered.

From the form of equations (22.4) it follows that a path is uniquely determined by a point P_0 of coördinates x^i_0 and a direction at P_0. In fact, if we put

$$\text{(22.5)} \qquad \xi^i = \left(\frac{dx^i}{ds} \right)_0,$$

where a subscript zero indicates the value at P_0, we have from (22.4)

* 1926, 1, p. 50.

$$(22.6)\qquad \begin{aligned} x^i = x^i_0 + \xi^i s - \frac{1}{2}(\Gamma^i_{jk})_0\, \xi^j \xi^k s^2 + \cdots \\ \cdots + \frac{1}{n!}\left(\frac{d^n x^i}{ds^n}\right)_0 s^n + \cdots, \end{aligned}$$

the coefficients of s^3 and higher powers of s being obtained from the equations which result from (22.4) by differentiation and reduction by means of (22.4). Thus we have

$$\begin{aligned} \frac{d^3 x^i}{ds^3} + \Gamma^i_{jkl} \frac{dx^j}{ds}\frac{dx^k}{ds}\frac{dx^l}{ds} &= 0, \\ \frac{d^4 x^i}{ds^4} + \Gamma^i_{jklm} \frac{dx^j}{ds}\frac{dx^k}{ds}\frac{dx^l}{ds}\frac{dx^m}{ds} &= 0, \\ \cdots\cdots\cdots\cdots\cdots\cdots\cdots \end{aligned}$$

where

$$(22.7)\qquad \begin{aligned} \Gamma^i_{jkl} &= \frac{1}{3} P\left(\frac{\partial}{\partial x^l}\Gamma^i_{jk} - \Gamma^i_{rk}\Gamma^r_{jl} - \Gamma^i_{jr}\Gamma^r_{kl}\right) \\ &= \frac{1}{3} P\left(\frac{\partial}{\partial x^l}\Gamma^i_{jk} - 2\,\Gamma^i_{jr}\Gamma^r_{kl}\right), \end{aligned}$$

and in general

$$(22.8)\qquad \Gamma^i_{jkl\cdots mp} = \frac{1}{N} P\left[\frac{\partial \Gamma^i_{jkl\cdots m}}{\partial x^p} - (N-1)\,\Gamma^i_{rk\cdots m}\Gamma^r_{jp}\right],$$

P before an expression indicating the sum of terms obtained by permuting the subscripts cyclically and N denotes the number of subscripts. Hence we have in place of (22.6)

$$(22.9)\qquad x^i = x^i_0 + \xi^i s - \frac{1}{2}(\Gamma^i_{jk})_0\, \xi^j \xi^k s^2 - \frac{1}{3!}(\Gamma^i_{jkl})_0\, \xi^j \xi^k \xi^l s^3 + \cdots.$$

The domain of convergence of these series depends evidently upon the expressions for Γ^i_{jk} and the values of ξ^i. However for sufficiently small values of s they define a path, that is, an integral curve of equations (22.4).*

23. Normal coördinates. In § 20 we saw that for a given symmetric connection there can be chosen coördinate

* These results are an immediate generalization of a similar treatment for geodesics in Riemannian geometry, 1926, 1, p. 52; cf. *Veblen* and *T. Y. Thomas*, 1923, 1, p. 560.

systems for any point so that at the point the coefficients Γ^i_{jk} are zero. In this section we wish to establish the existence of a class of coördinate systems possessing this property which are a generalization of Riemannian coördinates in a general Riemannian space.* They were considered first by Veblen,† who has called them *normal coördinates.*

Let C be a path through a point P_0, and s an affine parameter of the path which is zero at P_0; then the constants ξ^i are uniquely defined by (22.5). To each point of C we assign coördinates y^i by the equations

$$y^i = \xi^i s. \tag{23.1}$$

Since equations (22.9) define the path in the x's, between the x's and y's at points of C we have the relations

$$x^i = x^i_0 + y^i - \frac{1}{2}(\Gamma^i_{jk})_0 y^j y^k - \frac{1}{3!}(\Gamma^i_{jkl})_0 y^j y^k y^l + \cdots. \tag{23.2}$$

If we assign coördinates in this manner to the points on all paths through P_0 in a domain, such that no two paths meet again within it, we have a coördinate system y^i in the domain. Moreover, equations (23.2) are the same for all paths and consequently are the equations of the transformation of coördinates, P_0 being the origin for the y's. Since the jacobian $\left|\frac{\partial x^i}{\partial y^j}\right|$ of these equations is different from zero at P_0, the series can be inverted, and we have

$$y^i = x^i - x^i_0 + F^i(x^1 - x^1_0, \cdots, x^n - x^n_0), \tag{23.3}$$

where F^i are series in the second and higher powers of $(x^j - x^j_0)$ for $j = 1, \cdots, n$. Comparing (23.2) with (20.1) we see that the y's are a particular type of geodesic coördinates; we call them *normal* coördinates.

From the definition of the y's it follows that (23.1) are the equations in finite form in the y's of the paths through P_0.

* 1926, 1, p. 53.

† 1922, 5. p. 193; also *Veblen* and *T. Y. Thomas*, 1923, 1, pp. 562–566.

Consequently the equations of the paths through the origin of a normal coördinate system have the same form in these coördinates as the equations of straight lines in euclidean space in cartesian coördinates.

If we denote by C^i_{jk} the coefficients of the connection in the y's, the equations of the paths in this coördinate system are

$$\frac{d^2 y^i}{ds^2} + C^i_{jk} \frac{dy^j}{ds} \frac{dy^k}{ds} = 0. \tag{23.4}$$

Since these must be satisfied by (23.1), we must have

$$C^i_{jk} \xi^j \xi^k = 0, \tag{23.5}$$

and on multiplication by s^2

$$C^i_{jk} y^j y^k = 0, \tag{23.6}$$

which equations hold throughout the domain. Conversely, if these conditions are satisfied, equations (23.4) are satisfied by (23.1) and consequently the y's are normal coördinates.

When we apply to (23.4) considerations similar to those applied to (22.4) which led to (22.9), we obtain

$$\begin{aligned} y^i = \xi^i s - \frac{1}{2} (C^i_{jk})_0 \xi^j \xi^k s^2 - \frac{1}{3!} (C^i_{jkl})_0 \xi^j \xi^k \xi^l s^3 - \dots \\ - \frac{1}{p!} (C^i_{j_1 \dots j_p})_0 \xi^{j_1} \dots \xi^{j_p} s^p - \dots, \end{aligned} \tag{23.7}$$

where $C^i_{j_1 \dots j_p}$ are defined by equations of the form (22.8). Since these expressions must be equivalent to (23.1), we must have

$$(C^i_{jk})_0 = 0 \tag{23.8}$$

and

$$(C^i_{j_1 \dots j_p})_0 = 0 \tag{23.9}$$

for all values of p. Equations (23.8) follow also from (23.5), since the ξ's are arbitrary.

From (23.8) also it follows that normal coördinates are a particular class of geodesic coördinates (§ 20).

If instead of a general coördinate system x^i we take any other coördinate system x'^i and proceed as above and denote by y'^i the normal coördinates thus obtained, we have in place of (23.1)

$$y'^i = \left(\frac{dx'^i}{ds}\right)_0 s$$

for the equations of the paths. Since

$$\left(\frac{dx'^i}{ds}\right)_0 = \left(\frac{\partial x'^i}{\partial x^j}\frac{dx^j}{ds}\right)_0 = a^i_j \xi^j, \tag{23.10}$$

where the a's are constants, we have

$$y'^i = a^i_j y^j. \tag{23.11}$$

Hence we have:

When the coördinates x^i of a space are subjected to an arbitrary analytic transformation, the normal coördinates determined by the x's and a point undergo a linear homogeneous transformation with constant coefficients.

From the definition (23.10) of the a's it follows that when a transformation (23.11) of the normal coördinates is given, corresponding analytic transformations of the x's exist but are not uniquely defined.

From the form of (23.11) it follows that normal coördinates are fundamental in the affine geometry in the neighborhood of a point.

If we differentiate equation (23.6) with respect to s along any path, make use of (23.1) and multiply the resulting equation by s, we obtain

$$C^i_{jkl}\, y^j y^k y^l = 0, \tag{23.12}$$

where

$$C^i_{jkl} = \frac{1}{3} P\left(\frac{\partial C^i_{jk}}{\partial y^l}\right), \tag{23.13}$$

P indicating the sum of terms obtained by permuting j, k and l cyclically. Proceeding with (23.12) as was done with (23.6), we get a sequence of identities of the form

$$C^i_{j_1 \cdots j_r} y^{j_1} \cdots y^{j_r} = 0, \tag{23.14}$$

where

$$C^i_{j_1 \cdots j_r} = \frac{1}{r} P\left(\frac{\partial C^i_{j_1 \cdots j_{r-1}}}{\partial y^{j_r}}\right). \tag{23.15}$$

As thus defined the C's are symmetric in the subscripts and they are the functions in the normal coördinates y^i for which the corresponding functions in the x's are given by (22.8). From (23.13) and (23.15) we obtain

$$C^i_{jklm} = \frac{2}{3 \cdot 4} S\left(\frac{\partial^2 C^i_{jk}}{\partial y^l \partial y^m}\right),$$

S indicating the sum of the six terms obtained by the permutations of the subscripts, j, k, l, m, which do not yield equivalent terms. In general, we have

$$C^i_{j_1 \cdots j_r} = \frac{2}{r(r-1)} S\left(\frac{\partial^{r-2} C^i_{j_1 j_2}}{\partial y^{j_3} \cdots \partial y^{j_r}}\right); \tag{23.16}$$

in this case S indicates the sum of $r(r-1)/2$ different terms.

If y^i are a system of normal coördinates with a point P as origin and y'^i are the normal coördinates corresponding to the y's with the point P', of coördinates dy^i, as origin, we have from (23.2)

$$\begin{aligned} y^i = {} & dy^i + y'^i - \frac{1}{2} \overline{C}^i_{jk} y'^j y'^k \\ & - \frac{1}{3!} \overline{C}^i_{jkl} y'^j y'^k y'^l + \cdots, \end{aligned} \tag{23.17}$$

where $\overline{C}^i_{jk}$ and so forth are the values of the corresponding C's for the y's at P'. Because of (23.8) we have

$$\overline{C}^i_{jk} = \left(\frac{\partial C^i_{jk}}{\partial y^l}\right)_0 dy^l + \frac{1}{2}\left(\frac{\partial^2 C^i_{jk}}{\partial y^l \partial y^m}\right)_0 dy^l dy^m + \cdots.$$

24. Curvature of a curve. Let C be any curve in a V_n, not a path, the coördinates x^i being expressed in terms of a general parameter t. The equations

$$\frac{d^2 x^i}{dt^2} + L^i_{jk} \frac{dx^j}{dt} \frac{dx^k}{dt} = \bar{\mu}^i \tag{24.1}$$

define a contravariant vector $\bar{\mu}^i$. If we change the parameter t, we get in general a new vector, which is of the pencil determined by $\bar{\mu}^i$ and $\frac{dx^i}{dt}$. We single out one of these vectors by choosing for the parameter an affine parameter s of the path tangent to C at a given point and we choose s so that $s = 0$ at P. Accordingly we write

$$\frac{d^2 x^i}{ds^2} + L^i_{jk} \frac{dx^j}{ds} \frac{dx^k}{ds} = \mu^i,$$

which in fact are equivalent to

$$\frac{d^2 x^i}{ds^2} + \Gamma^i_{jk} \frac{dx^j}{ds} \frac{dx^k}{ds} = \mu^i, \tag{24.2}$$

that is, the vector μ^i is not affected by the choice of the tensor Ω^i_{jk}. Since s is determined to within a constant factor (§ 22), the same is true of μ^i.

If we take for the x's a set of normal coördinates with origin at P, these equations reduce to

$$\frac{d^2 x^i}{ds^2} = \mu^i \tag{24.3}$$

These equations are an evident generalization of those in euclidean three-space which define the first curvature vector of the curve at P. Accordingly we call μ^i defined by (24.2) in general coördinates the *first curvature vector* of the curve at P.

From (24.3) we have for C

$$x^i = \left(\frac{dx^i}{ds}\right)_0 s + \frac{1}{2} (\mu^i)_0 s^2 + \cdots, \tag{24.4}$$

and the equations of the path tangent to C at P are

$$\bar{x}^i = \left(\frac{dx^i}{ds}\right)_0 s.$$

Consequently the values of μ^i at P determine the departure of the curve from the path at a point near P.

It is readily shown that

*The surface formed by the paths through a point of a curve in the pencil of directions determined by the tangent and first curvature vector to the curve at the point osculates the curve.**

25. Extension of the theorem of Fermi to symmetric connections. The following theorem was proved for Riemannian connections by Fermi† and in this section is established for symmetric connections:

For a space with a symmetric connection it is possible to choose a coördinate system with respect to which the coefficients Γ^i_{jk} are zero at all points of a curve, or of a portion of it.

Suppose that the curve C is defined by $x^i = \varphi^i(t)$ and that at a point P_0 of it we take $n-1$ independent vectors $\lambda^i_{(\alpha)}$ for $\alpha = 1, \cdots, n-1$, which also are independent of the tangent to the curve, that is, at P_0 the determinant

$$\Delta = \begin{vmatrix} \lambda^1_{(1)} & \cdots & \lambda^n_{(1)} \\ \cdot & \cdot & \cdot \\ \cdot & \cdot & \cdot \\ \lambda^1_{(n-1)} & \cdots & \lambda^n_{(n-1)} \\ \varphi^{1\prime} & \cdots & \varphi^{n\prime} \end{vmatrix} \tag{25.1}$$

is different from zero, primes indicating differentation with respect to t. From these vectors we obtain $n-1$ families of vectors $\lambda^i_{(\alpha)}$ by parallel displacement along C. It follows from continuity considerations that there is a portion R of C about P_0 for which $\Delta \neq 0$. At P_0 the components of any vector depending upon the given $n-1$ vectors are of the form

$$\lambda^i = A^\alpha \lambda^i_{(\alpha)} \quad (\alpha = 1, \cdots, n-1). \tag{25.2}$$

If this vector undergoes parallel displacement along C, we get a family of vectors whose components at each point are

* Cf. 1926, 1, p. 62.

† 1922, 7; the method followed is an adaptation of a proof of the theorem for Riemannian connections given by *Levi-Civita*, 1926, 4.

given by (25.2) in which the A's remain constant. Since any point can be taken as P_0, the components of any vector at any point in the $(n-1)$-fold of vectors at the point can be expressed in the form (25.2).

If we put

$$\lambda^i_{(n)} = f(t)\varphi^{i'} + \lambda^i_{(\alpha)} f^\alpha(t) \tag{25.3}$$

and express the condition that the functions $\lambda^i_{(n)}$ are a solution of (7.1), we have

$$f\bar{\mu}^i + f'\varphi^{i'} + \lambda^i_{(\alpha)} f^{\alpha\prime} = 0, \tag{25.4}$$

where $\bar{\mu}^i$ are defined by (24.1). In the region R functions $a(t)$ and $a^\alpha(t)$ can be determined such that

$$\bar{\mu}^i = a\varphi^{i'} + a^\alpha \lambda^i_{(\alpha)}, \tag{25.5}$$

since $\Delta \neq 0$, and the functions f and f^α are determined by the quadratures

$$f = ce^{-\int a dt}, \quad f^\alpha = -\int f a^\alpha dt + c^\alpha, \tag{25.6}$$

where the c's are constants. If in particular the given curve is a path $a^\alpha = 0$ for $\alpha = 1, \cdots, n-1$.

Consider at any point P of C the $(n-1)$-fold of vectors defined by (25.2) and the paths of the space through P in these directions. The locus of these paths is a V_{n-1}. The equations of any one of these paths are

$$x^i = \varphi^i(t) + A^\alpha \lambda^i_{(\alpha)} s - \frac{1}{2}(\Gamma^i_{jk})_P A^\alpha \lambda^j_{(\alpha)} A^\beta \lambda^k_{(\beta)} s^2 + \cdots, \tag{25.7}$$

where t is the value at P of the parameter along C, $(\Gamma^i_{jk})_P$ are evaluated at P and s, the affine parameter of the path, is chosen (§ 22) so that at P we have $\frac{dx^i}{ds} = A^\alpha \lambda^i_{(\alpha)}$. A new set of coördinates y^i is assigned to each point of C by means of equations

$$y^i = \psi^i(t),$$

where the ψ's are any continuous functions of t. In like manner along each path through P we associate coördinates y^i with each point by means of the equations

$$(25.8) \qquad y^\alpha = \psi^\alpha(t) + A^\alpha s, \quad y^n = \psi^n(t),$$

where the A's are the constants in (25.2) which determine the direction of the path at P; thus $s = 0$ at P. We assume that $\psi^{n'} \neq 0$, so that the last of (25.8) can be written

$$(25.9) \qquad t = \theta(y^n).$$

We eliminate the A's from (25.7) by means of (25.8) and replace t by its expression (25.9). This gives equations of the form

$$(25.10) \qquad \begin{aligned} x^i &= F^i(y^n) + \lambda^i_{(\alpha)}(y^\alpha - \overline{\psi}^\alpha) \\ &\quad - \frac{1}{2}(\Gamma^i_{jk})\lambda^j_{(\alpha)}\lambda^k_{(\beta)}(y^\alpha - \overline{\psi}^\alpha)(y^\beta - \overline{\psi}^\beta) + \cdots, \end{aligned}$$

where the λ's, $\overline{\psi}$'s and (Γ^i_{jk}) are functions of y^n. If this process is followed out at each point and for each direction, we have coördinates y^i associated with every point of the family of V_{n-1}'s as defined, and equations (25.10) are of the same form for all the points. At points of C, that is where $y^\alpha = \overline{\psi}^\alpha$, the jacobian of (25.10) is reducible to $\theta\Delta$ or $\Delta/\psi^{n'}$, where Δ is given by (25.1). Hence for a domain of the space in the neighborhood of the portion R of the curve for which $\Delta \neq 0$, equations (25.10) define a transformation of coördinates.

If we denote by $\overline{\lambda}^i_{(\alpha)}$ the components in the y's of the vectors $\lambda^i_{(\alpha)}$, we have at points of C, by means of (25.10),

$$\lambda^i_{(\alpha)} = \overline{\lambda}^j_{(\alpha)} \frac{\partial x^i}{\partial y^j} = \overline{\lambda}^\beta_{(\alpha)}\lambda^i_{(\beta)} + \overline{\lambda}^n_{(\alpha)}\left(\frac{dF^i}{dy^n} - \lambda^i_{(\beta)}\frac{d\overline{\psi}^\beta}{dy^n}\right).$$

When these equations are written in the form

$$\lambda^i_{(\beta)}(\overline{\lambda}^\beta_{(\alpha)} - \delta^\beta_\alpha) + \overline{\lambda}^n_{(\alpha)}(\varphi^{i'} - \lambda^i_{(\beta)}\psi^{\beta'})\frac{1}{\psi^{n'}} = 0,$$

we have that

$$(25.11) \qquad \bar{\lambda}^i_{(\alpha)} = \delta^i_\alpha .$$

In order that a family of vectors $\lambda^i_{(n)}$ may have the components

$$(25.12) \qquad \bar{\lambda}^i_{(n)} = \delta^i_n$$

in the y's, it is necessary that

$$\lambda^i_{(n)} = \bar{\lambda}^j_{(n)} \frac{\partial x^i}{\partial y^j} = \frac{\partial x^i}{\partial y^n} = (\varphi^{i\prime} - \lambda^i_{(\alpha)} \psi^{\alpha\prime}) \frac{1}{\psi^{n\prime}} .$$

Comparing this result with (25.3), (25.4) and (25.6), we see that, if in the above definition of the y's we take

$$\psi^{n\prime} = \frac{1}{f}, \qquad \psi^{\alpha\prime} = \psi^{n\prime} f^\alpha,$$

where the f's are given by (25.6), then the components in the y's of the vectors (25.3) will have the values (25.12), and will be parallel along the curve because of (25.6).

From (25.10) we have at points of C

$$\frac{\partial x^i}{\partial y^\alpha} = \lambda^i_{(\alpha)} ,$$

$$\frac{\partial^2 x^i}{\partial y^\alpha \partial y^\beta} = -(\Gamma^i_{jk}) \lambda^j_{(\alpha)} \lambda^k_{(\beta)} = -(\Gamma^i_{jk}) \frac{\partial x^j}{\partial y^\alpha} \frac{\partial x^k}{\partial y^\beta} .$$

Hence from equations of the form (5.6), we have in the y's

$$(25.13) \qquad \Gamma^i_{\alpha\beta} = 0 \qquad (\alpha, \beta = 1, \cdots, n-1;\ i = 1, \cdots, n).$$

Since by hypothesis the vectors of components (25.11) and (25.12) are parallel along C, we have

$$\bar{\Gamma}^i_{jl} \frac{d y^l}{d t} = 0 .$$

If $j = 1, \cdots, n-1$, we have in consequence of (25.13) and $\psi^{n\prime} \neq 0$, that $\bar{\Gamma}^i_{\alpha n} = 0$, and then for $j = n$ that $\bar{\Gamma}^i_{nn} = 0$. Consequently we have established the theorem.

We observe that only the first three terms of (25.10) have entered in the above discussion. Consequently, any expressions differing from (25.10) in terms of the third and higher orders of $(y^\alpha - \overline{\psi}^\alpha)$ define transformations of the desired type.

26. Normal tensors. Because of the conditions (23.8), when the functions C^i_{jk} are developed in powers of the y's, we have

$$C^i_{jk} = A^i_{jkl}\, y^l + \frac{1}{2} A^i_{jkl_1 l_2}\, y^{l_1} y^{l_2} + \cdots \tag{26.1}$$
$$+ \frac{1}{r!} A^i_{jkl_1 \cdots l_r}\, y^{l_1} \cdots y^{l_r} + \cdots,$$

where

$$A^i_{jkl_1 \cdots l_r} = \left(\frac{\partial^r C^i_{jk}}{\partial y^{l_1} \cdots \partial y^{l_r}}\right)_0. \tag{26.2}$$

From the equations of transformation (23.2) we have

$$\left(\frac{\partial x^i}{\partial y^j}\right)_0 = \delta^i_j, \qquad \left(\frac{\partial^2 x^i}{\partial y^j \partial y^k}\right)_0 = -(\Gamma^i_{jk})_0, \tag{26.3}$$
$$\left(\frac{\partial^r x^i}{\partial y^{j_1} \cdots \partial y^{j_r}}\right)_0 = -(\Gamma^i_{j_1 \cdots j_r})_0.$$

In consequence of the first set of these equations it follows that, if at the origin the numbers (26.2) are taken as the components of a tensor in the y's, the components of the same tensor in the x's have the same values.

If we take any other coördinate system x'^i and the corresponding normal coördinates y'^i with the same origin as above, we have a new set of constants defined by

$$A'^i_{jkl_1 \cdots l_r} = \left(\frac{\partial^r C'^i_{jk}}{\partial y'^{l_1} \cdots \partial y'^{l_r}}\right)_0. \tag{26.4}$$

We will show that the A's and A''s are the components of the same tensor in the y's and y''s respectively, and also in the x's and x''s respectively. From (23.11) it follows that

$$\frac{\partial y'^i}{\partial y^j} = a^i_j, \qquad \frac{\partial^2 y'^i}{\partial y^j \partial y^k} = 0. \tag{26.5}$$

Consequently the functions C^i_{jk} and C'^i_{jk} satisfy the relations [cf. (5.6)]

$$C^\alpha_{jk}\frac{\partial y'^i}{\partial y^\alpha} = C'^i_{\beta\gamma}\frac{\partial y'^\beta}{\partial y^j}\frac{\partial y'^\gamma}{\partial y^k}. \tag{26.6}$$

Since a^i_j in (26.5) are constants, we have from (26.6) by differentiation

$$\begin{aligned}\frac{\partial^r C^\alpha_{jk}}{\partial y^{l_1}\dots\partial y^{l_r}}\frac{\partial y'^i}{\partial y^\alpha}\\ = \frac{\partial^r C'^i_{\beta\gamma}}{\partial y'^{\sigma_1}\dots\partial y'^{\sigma_r}}\frac{\partial y'^\beta}{\partial y^j}\frac{\partial y'^\gamma}{\partial y^k}\frac{\partial y'^{\sigma_1}}{\partial y^{l_1}}\cdots\frac{\partial y'^{\sigma_r}}{\partial y^{l_r}}.\end{aligned} \tag{26.7}$$

At the origin of the two sets of normal coördinates

$$\left(\frac{\partial x'^i}{\partial x^j}\right)_0 = \left(\frac{\partial x'^i}{\partial y'^\beta}\frac{\partial y'^\beta}{\partial y^\alpha}\frac{\partial y^\alpha}{\partial x^j}\right)_0 = \frac{\partial y'^i}{\partial y^j}, \tag{26.8}$$

in consequence of (26.3) and similar equations. Consequently at the origin we have from (26.2), (26.4), (26.7) and (26.8)

$$A^\alpha_{jkl_1\cdots l_r}\frac{\partial x'^i}{\partial x^\alpha} = A'^i_{\beta\gamma\sigma_1\cdots\sigma_r}\frac{\partial x'^\beta}{\partial x^j}\frac{\partial x'^\gamma}{\partial x^k}\frac{\partial x'^{\sigma_1}}{\partial x^{l_1}}\cdots\frac{\partial x'^{\sigma}}{\partial x^{l_r}}.$$

Hence if at each point in space we obtain the numbers $A^\alpha_{jkl_1\cdots l_r}$ and $A'^i_{\beta\gamma\sigma_1\cdots\sigma_r}$ by the processes (26.2) and (26.4), these are the components in the x's and x''s of a tensor. Being defined at each point of space, the A's and A''s may be regarded as functions of the x's and x''s. In fact, we shall show presently what the functional forms of certain of them are. Following Veblen and T. Y. Thomas,* who have developed this theory, we call them *normal* tensors.

If we differentiate the equations

$$C^\alpha_{jk}\frac{\partial x^i}{\partial y^\alpha} = \frac{\partial^2 x^i}{\partial y^j\partial y^k} + \Gamma^i_{\beta\gamma}\frac{\partial x^\beta}{\partial y^j}\frac{\partial x^\gamma}{\partial y^k} \tag{26.9}$$

* 1923, 1, p. 567.

with respect to y^l and make use of (26.3), we have at the origin

$$A^i_{jkl} = \frac{\partial \Gamma^i_{jk}}{\partial x^l} - \Gamma^i_{jkl} - \Gamma^i_{\beta k} \Gamma^\beta_{jl} - \Gamma^i_{j\beta} \Gamma^\beta_{kl}. \tag{26.10}$$

Since any point may be taken as the origin, equations (26.10) define A^i_{jkl} throughout space. If we differentiate (26.9) with respect to y^l and y^m, and proceed as above, we obtain

$$\begin{aligned} A^i_{jklm} = {} & \frac{\partial^2 \Gamma^i_{jk}}{\partial x^l \partial x^m} - \Gamma^i_{jklm} - \frac{\partial \Gamma^i_{jh}}{\partial x^l} \Gamma^h_{km} - \frac{\partial \Gamma^i_{hk}}{\partial x^l} \Gamma^h_{jm} \\ & - \frac{\partial \Gamma^i_{jh}}{\partial x^m} \Gamma^h_{kl} - \frac{\partial \Gamma^i_{hk}}{\partial x^m} \Gamma^h_{jl} - \frac{\partial \Gamma^i_{jk}}{\partial x^h} \Gamma^h_{lm} - \Gamma^i_{jh} \Gamma^h_{klm} \\ & - \Gamma^i_{hk} \Gamma^h_{jlm} + A^h_{jkl} \Gamma^i_{hm} + A^h_{jkm} \Gamma^i_{hl} \\ & + \Gamma^i_{hr} (\Gamma^h_{jl} \Gamma^r_{km} + \Gamma^h_{jm} \Gamma^r_{kl}). \end{aligned} \tag{26.11}$$

By continuing this process the components of a normal tensor of any order can be obtained.

From equations (26.2) it follows that the components $A^i_{jkl_1 \cdots l_r}$ are symmetric in j and k and in the last r indices. In consequence of (23.6) we have that, if C^i_{jk} as given by (26.1) is multiplied by $y^j y^k$ and j and k are summed, each term on the right must be zero, that is

$$P(A^i_{jkl_1 \cdots l_r}) = 0, \tag{26.12}$$

where P indicates the sum of all the terms obtained by permuting the indices. However, because of the above observation concerning symmetry, this equation can be replaced by

$$S(A^i_{jkl_1 \cdots l_r}) = 0, \tag{26.13}$$

where S indicates the sum of the $(r+1)(r+2)/2$ terms obtained by the permutations of the subscripts which do not yield equivalent terms. Thus for $r = 1$ and $r = 2$ we have respectively

$$A^i_{jkl} + A^i_{klj} + A^i_{ljk} = 0, \tag{26.14}$$

$$A^i_{jklm} + A^i_{jlmk} + A^i_{jmkl} + A^i_{kljm} + A^i_{kmjl} + A^i_{lmjk} = 0. \tag{26.15}$$

Because of (23.9) the functions $C^i_{j_1\cdots j_r}$, as defined by (23.15), are expressible as power series in y's of the form

$$(26.16)\quad C^i_{j_1\cdots j_r} = A^i_{(j_1\cdots j_r)l}\, y^l + \frac{1}{2}\, A^i_{(j_1\cdots j_r)l_1 l_2}\, y^{l_1} y^{l_2} + \cdots,$$

where because of (23.16) and (26.2)

$$(26.17)\quad \begin{aligned} A^i_{(j_1\cdots j_r)l_1\cdots l_s} &= \left(\frac{\partial^s C^i_{j_1\cdots j_r}}{\partial y^{l_1}\cdots\partial y^{l_s}}\right)_0 \\ &= \frac{2}{r(r-1)}\, \overline{S}\,(A^i_{j_1\cdots j_r l_1\cdots l_s}), \end{aligned}$$

the functions $A^i_{j_1\cdots j_r l_1\cdots l_s}$ being components of a normal tensor and $\overline{S}$ indicating the sum of $r(r-1)/2$ different terms obtained by the permutation of the indices $j_1, \cdots, j_r$. From these results it follows that $A^i_{(j_1\cdots j_r)l_1\cdots l_s}$ are components of a tensor in the x's, symmetric in the subscripts $j_1, \cdots, j_r$ and in the subscripts $l_1, \cdots, l_s$. When we apply to (26.16) reasoning similar to that which led to (26.12) from (26.1), we have

$$(26.18)\qquad P(A^i_{(j_1\cdots j_r)l_1\cdots l_s}) = 0,$$

giving identities connecting the components of these tensors.

Since Γ^i_{jkl}, as defined by (22.7), are symmetric in k and l, it follows from (5.4) and (26.10) that

$$(26.19)\qquad B^i_{jkl} = A^i_{jlk} - A^i_{jkl}.$$

The identities (21.3) follow directly from this result; and (21.4) because the A's are symmetric in the first two indices. From (26.14) and (26.19) we have

$$(26.20)\quad A^i_{jkl} = \frac{1}{3}(2B^i_{jlk} + B^i_{lkj}) = \frac{1}{3}(B^i_{jlk} + B^i_{klj}).$$

From (26.19) by covariant differentiation we obtain

$$B^i_{jkl,m} = A^i_{jlk,m} - A^i_{jkl,m}.$$

In normal coördinates y^i we have because of (26.10) and (26.17)

$$A^i_{jkl,m} = A^i_{jklm} - A^i_{(jkl)m}.$$

Consequently we have

$$B^i_{jkl,m} = A^i_{jlkm} - A^i_{jklm} \tag{26.21}$$

in any coördinate system.

In like manner we obtain

$$\begin{aligned} B^i_{jkl,m_1m_2} = {} & A^i_{jlkm_1m_2} - A^i_{jklm_1m_2} + A^i_{\beta km_1} A^\beta_{jlm_2} - A^i_{\beta lm_1} A^\beta_{jkm_2} \\ & + A^\beta_{jlm_1} A^i_{\beta km_2} - A^\beta_{jkm_1} A^i_{\beta lm_2} + A^i_{hm_1m_2} (A^h_{jlk} - A^h_{jkl}) \\ & + A^h_{jm_1m_2} (A^i_{khl} - A^i_{hlk}) + A^h_{km_1m_2} (A^i_{jhl} - A^i_{jlh}) \\ & + A^h_{lm_1m_2} (A^i_{jkh} - A^i_{jhk}). \end{aligned}$$

By this method we are able to express any covariant derivative of the curvature tensor in terms of normal tensors.*

27. Extensions of a tensor. In this section we define a process of obtaining tensors of higher order by differentiation, suggested by the method of obtaining normal tensors.

Consider a tensor of components $T^{i_1\cdots i_r}_{j_1\cdots j_s}$ and $T'^{i_1\cdots i_r}_{j_1\cdots j_s}$ in general coördinates x^i and x'^i, and let $t^{i_1\cdots i_r}_{j_1\cdots j_s}$ and $t'^{i_1\cdots i_r}_{j_1\cdots j_s}$ denote the components of this tensor in the normal coördinates y^i and y'^i, corresponding to x^i and x'^i respectively, for the same origin P. Accordingly we have

$$t'^{i_1\cdots i_r}_{j_1\cdots j_s} = t^{\alpha_1\cdots\alpha_r}_{\beta_1\cdots\beta_s} \frac{\partial y'^{i_1}}{\partial y^{\alpha_1}} \cdots \frac{\partial y'^{i_r}}{\partial y^{\alpha_r}} \frac{\partial y^{\beta_1}}{\partial y'^{j_1}} \cdots \frac{\partial y^{\beta_s}}{\partial y'^{j_s}}.$$

Since the quantities $\frac{\partial y'^i}{\partial y^\alpha}$ and $\frac{\partial y^\alpha}{\partial y'^i}$ are constants (§ 23), we have from the preceding equations on differentiating with respect to $y'^{k_1}, \ldots, y'^{k_p}$,

* For further developments concerning normal tensors see *Veblen* and *T. Y. Thomas*, 1923, 1, pp. 576–580.

$$(27.1)\quad \frac{\partial^p t'^{i_1\cdots i_r}_{j_1\cdots j_s}}{\partial y'^{k_1}\cdots\partial y'^{k_p}} = \frac{\partial^p t^{\alpha_1\cdots\alpha_r}_{\beta_1\cdots\beta_s}}{\partial y^{\gamma_1}\cdots\partial y^{\gamma_p}} \times \frac{\partial y'^{i_1}}{\partial y^{\alpha_1}}\cdots\frac{\partial y^{\beta_s}}{\partial y'^{j_s}}\frac{\partial y^{\gamma_1}}{\partial y'^{k_1}}\cdots\frac{\partial y^{\gamma_p}}{\partial y'^{k_p}}.$$

If we put

$$(27.2)\quad T^{\alpha_1\cdots\alpha_r}_{\beta_1\cdots\beta_s;\gamma_1\cdots\gamma_p} = \left(\frac{\partial^p t^{\alpha_1\cdots\alpha_r}_{\beta_1\cdots\beta_s}}{\partial y^{\gamma_1}\cdots\partial y^{\gamma_p}}\right)_0$$

and similarly for the t''s then at P in consequence of (27.1) and (26.8), we have

$$T'^{i_1\cdots i_r}_{j_1\cdots j_s;k_1\cdots k_p} = T^{\alpha_1\cdots\alpha_r}_{\beta_1\cdots\beta_s;\gamma_1\cdots\gamma_p}\frac{\partial x'^{i_1}}{\partial x^{\alpha_1}}\cdots\frac{\partial x'^{i_r}}{\partial x^{\alpha_r}}\cdots\frac{\partial x^{\gamma_p}}{\partial x'^{k_p}}.$$

Hence at P the numbers so defined are the components of a tensor in the coördinate systems x^i and x'^i. Since P is any point, we have thus at every point the components of a tensor in the two coördinate systems, and thus the T's and T''s of (27.2) are functions of the x's and x''s, as in the case of the normal tensors. Following Veblen and T. Y. Thomas* we call them *extensions* of the pth *order*, when there are p additional subscripts as indicated in (27.2), From (27.2) it is seen that an extension is symmetric in the subscripts indicating differentiation, whereas this is not the case for covariant derivatives.

When $p = 1$, the right-hand member of (27.2) is equal to the first covariant derivative of $t^{\alpha_1\cdots\alpha_r}_{\beta_1\cdots\beta_s}$ at P and consequently the left-hand member is the first covariant derivative of $T^{\alpha_1\cdots\alpha_r}_{\beta_1\cdots\alpha_s}$ in the x's, evaluated at P. However, when $p > 1$, the pth extension is not equal to the pth covariant derivative, since the second and higher covariant derivatives involve derivatives of the coefficients $C^i_{\alpha\beta}$ which are not zero at P.

* 1923, 1, p. 572. We use a different notation in that a comma followed by p subscripts indicates the pth covariant derivative and a semi-colon followed by p subscripts the pth extension..

In order to obtain an expression for the second extension of $T^{\alpha_1 \cdots \alpha_r}_{\beta_1 \cdots \beta_s}$, we observe that at P the second covariant derivative of $t^{\alpha_1 \cdots \alpha_r}_{\beta_1 \cdots \beta_s}$ is given by [cf. (6.1)]

$$t^{\alpha_1 \cdots \alpha_r}_{\beta_1 \cdots \beta_s, \gamma \delta} = \frac{\partial^2 t^{\alpha_1 \cdots \alpha_r}_{\beta_1 \cdots \beta_s}}{\partial y^\gamma \partial y^\delta} + \sum_i^{1, \cdots, r} t^{\alpha_1 \cdots \alpha_{i-1} \sigma \alpha_{i+1} \cdots \alpha_r}_{\beta_1 \cdots \beta_s} \frac{\partial C^{\alpha_i}_{\sigma \gamma}}{\partial y^\delta} - \sum_j^{1, \cdots, s} t^{\alpha_1 \cdots \alpha_r}_{\beta_1 \cdots \beta_{j-1} \tau \beta_{j+1} \cdots \beta_s} \frac{\partial C^\tau_{\beta_j \gamma}}{\partial y^\delta}.$$

From the preceding observation and those of § 26 we have in the x's at P

$$(27.3) \qquad T^{\alpha_1 \cdots \alpha_r}_{\beta_1 \cdots \beta_s, \gamma \delta} = T^{\alpha_1 \cdots \alpha_r}_{\beta_1 \cdots \beta_s; \gamma \delta} + \sum_i^{1, \cdots, r} T^{\alpha_1 \cdots \alpha_{i-1} \sigma \alpha_{i+1} \cdots \alpha_r}_{\beta_1 \cdots \beta_s} A^{\alpha_i}_{\sigma \gamma \delta} - \sum_j^{1, \cdots, s} T^{\alpha_1 \cdots \alpha_r}_{\beta_1 \cdots \beta_{j-1} \tau \beta_{j+1} \cdots \beta_s} A^\tau_{\beta_j \gamma \delta},$$

where the A's are normal tensors. Since P is any point, we have thus the general relation connecting covariant derivatives and extensions of the second order. Evidently this process may be extended to any order. In general, the difference between a pth covariant derivative and an extension of the pth order is expressible linearly in covariant derivatives of orders $p-2, p-3, \cdots, 1$ and the tensor itself, the coefficients being normal tensors covariant of orders $3, 4, \cdots, p+1$ respectively.*

The form of these expressions is not so important as the fact that there exist tensors whose components reduce to the derivatives alone at the origin of normal coördinates, as in (27.2). Moreover, we remark that both the covariant derivatives and the extensions are generalizations of ordinary derivatives in euclidean space referred to cartesian coördinates, since both expressions reduce to ordinary derivatives in this case.

28. The equivalence of symmetric connections. The question of whether a set of functions Γ^i_{jk} of coördi-

* Cf. *Veblen* and *T. Y. Thomas,* l. c., for a number of examples.

nates x^i and a set $\Gamma_{jk}^{\prime i}$ of coördinates $x^{\prime i}$ define the same symmetric connection reduces to the problem of determining whether the equations

$$(28.1) \qquad \frac{\partial^2 x^i}{\partial x^{\prime \beta} \partial x^{\prime \gamma}} + \Gamma_{jk}^i \frac{\partial x^j}{\partial x^{\prime \beta}} \frac{\partial x^k}{\partial x^{\prime \gamma}} = \Gamma_{\beta\gamma}^{\prime \alpha} \frac{\partial x^i}{\partial x^{\prime \alpha}}$$

admit a solution $x^i = \varphi^i(x^{\prime 1}, \cdots, x^{\prime n})$, such that the jacobian of the φ's is not zero. The conditions of integrability of these equations are (cf. § 21)

$$(28.2) \qquad B_{\beta\gamma\delta}^{\prime \alpha} u_\alpha^i = B_{jkl}^i u_\beta^j u_\gamma^k u_\delta^l,$$

where we have put

$$(28.3) \qquad \frac{\partial x^i}{\partial x^{\prime \alpha}} = u_\alpha^i.$$

When equations (28.2) are differentiated with respect to $x^{\prime \sigma_1}$, the resulting equations are reducible by means of (28.1) to

$$(28.4) \qquad B_{\beta\gamma\delta, \sigma_1}^{\prime \alpha} u_\alpha^i = B_{jkl, m_1}^i u_\beta^j \cdots u_{\sigma_1}^{m_1}.$$

Continuing this process, we obtain the infinite sequence of equations

$$
\begin{aligned}
(28.5) \qquad & B_{\beta\gamma\delta, \sigma_1\sigma_2}^{\prime \alpha} u_\alpha^i = B_{jkl, m_1 m_2}^i u_\beta^j \cdots u_{\sigma_2}^{m_2}, \\
& \cdot \quad \cdot \quad \cdot \quad \cdot \quad \cdot \quad \cdot \quad \cdot \quad \cdot \quad \cdot \quad \cdot \\
& B_{\beta\gamma\delta, \sigma_1 \cdots \sigma_r}^{\prime \alpha} u_\alpha^i = B_{jkl, m_1 \cdots m_r}^i u_\beta^j \cdots u_{\sigma_r}^{m_r}, \\
& \cdot \quad \cdot \quad \cdot \quad \cdot \quad \cdot \quad \cdot \quad \cdot \quad \cdot \quad \cdot \quad \cdot
\end{aligned}
$$

By means of (28.3) equations (28.1) can be written as

$$\frac{\partial u_\beta^i}{\partial x^{\prime \gamma}} = \Gamma_{\beta\gamma}^{\prime \alpha} u_\alpha^i - \Gamma_{jk}^i u_\beta^j u_\gamma^k.$$

These equations and (28.3) constitute a system of the form (8.1), such that the n^2 quantities u_α^i and the n quantities x^i are the functions θ^α, the $x^{\prime}$'s being the independent variables; consequently $M = n^2 + n$. The equations (28.2),

(28.4) and (28.5) are in this case the sets $F_1, F_2, \cdots$ of § 8. Accordingly we have:

A necessary and sufficient condition that two symmetric connections of coefficients Γ^i_{jk} *and* Γ'^i_{jk} *be equivalent is that there exist a positive integer N such that all sets of solutions of the equations of the first N sets of equations* (28.2), (28.4) *and* (28.5) *satisfy the* $(N+1)$*th set of these equations; if the number of independent equations of the first N sets is* n^2+n-p $(p \geqq 0)$, *the solution involves p arbitrary constants.*

From the considerations of § 8 it follows that, if the equations (28.2), (28.4) and (28.5) are consistent, then $N \leqq n^2+n$. We shall show that N and p are numerical invariants for a connected manifold. In fact, denote by $F'(\overset{i}{u}_\alpha, x^i, x'^i)$ the N sets of equations of the theorem in consequence of which the $(N+1)$th set is satisfied and let

$$x^i = \varphi^i_1(x'^1, \cdots, x'^n), \qquad \overset{i}{u}_\alpha = \frac{\partial \varphi^i_1}{\partial x'^\alpha}$$

be a solution of the problem. Let $x'^i = \psi^i(x''^1, \cdots, x''^n)$ define a transformation to a third set of variables x''^i. From these we have the equations

$$x^i = \varphi^i_2(x''^1, \cdots, x''^n) \tag{28.6}$$

defining the relations connecting the x's and x'''s. If we form the equations analogous to (28.2), (28.4) and (28.5) for the transformation (28.6) and denote by $F''(\overset{i}{\bar{u}}_\alpha, x^i, x''^i)$ the N sets of these equations in the x's and x'''s, it follows that these equations are satisfied by x^i, given by (28.6) and $\overset{i}{\bar{u}}_\beta = \overset{i}{u}_\alpha \dfrac{\partial \psi^\alpha}{\partial x''^\beta}$, and that the $(N+1)$th set also is satisfied. Moreover, the Nth set of these equations is not a consequence of the others; for, if it were, then by reversing the process we would have that the Nth set of the original group is a consequence of its predecessors in the sequence.

By the same argument it follows that the number n^2+n-p is an invariant for the manifold. Hence we have:

*The minimum positive integers N and p which enter in the determination of whether two given symmetric connections are equivalent are numerical invariants for the manifold.**

We say that a system of invariants is *complete*, when these invariants for two symmetric connections are sufficient to determine whether the two connections are equivalent. From the above results it follows that at most n^2+n+1 of the tensors $B^i_{jkl}, B^i_{jkl,m_1}, \cdots, B^i_{jkl,m_1\cdots m_r}, \cdots$ constitute a complete system of invariants for an affinely connected manifold; we have also that there exists a minimum positive integer N such that $N+1$ of the above tensors form a complete system. In consequence of the results of § 26 we have that $N+1$ of the sets of normal tensors $A^i_{jkl}, \cdots, A^i_{jklm_1\cdots m_r}, \cdots$ form a complete system also.

Christoffel† considered the problem of determining the necessary and sufficient conditions that two sets of functions g_{ij} and g'_{ij} of x^i and x'^i respectively be the components of the same tensor. The first condition is

$$g'_{\alpha\beta} = g_{ij}\frac{\partial x^i}{\partial x'^\alpha}\frac{\partial x^j}{\partial x'^\beta}. \tag{28.7}$$

When these equations are differentiated with respect to x'^γ, the resulting equations are equivalent to (1.5). These are of the form (28.1) and their conditions of integrability are given by (28.2), (28.4) and (28.5), where now the B's are the components of the Riemannian curvature tensor. Equations (28.7) are of the kind referred to in the latter part of § 8 as forming a set of conditions F_0. In this case, however, on differentiating (28.7), the resulting equations are satisfied because of (1.5), so that equations (28.2) are the set F_1 and so on. Accordingly the solution of the problem reduces to the consistency of (28.7), (28.2), (28.4) and (28.5) after the manner of the theorem of § 8, as Christoffel proved.

* Cf. *T. Y. Thomas* and *A. D. Michal*, 1927, 3.

† 1869, 1, p. 60.

For two asymmetric connections equations (2.1) reduce to (28.1) and

$$\Omega^i_{jk}\, u^j_\beta\, u^k_\gamma = \Omega'^\alpha_{\beta\gamma}\, u^i_\alpha . \tag{28.8}$$

Consequently in considering the question of equivalence of two such connections, equations (28.8) constitute the set F_0 of § 8. Then the set F_1 consists of equations (28.2) and

$$\Omega^i_{jk,m_1}\, u^j_\beta\, u^k_\gamma\, u^{m_1}_{\delta_1} = \Omega'^\alpha_{\beta\gamma,\delta_1}\, u^i_\alpha ; \tag{28.9}$$

and each other set F_r consists of the set in (28.4) or (28.5) involving the rth covariant derivatives of the B's and the equations of the form (28.9) involving the rth covariant derivatives of the Ω's. With this understanding the above theorem applies to this case.

29. Riemannian spaces. Flat spaces. When a space is Riemannian and g_{ij} are the components of the fundamental tensor, the Christoffel symbols of the second kind are the coefficients of a symmetric connection, as seen in § 1. In this case the following equations are satisfied identically:

$$g_{ij,k} = \frac{\partial g_{ij}}{\partial x^k} - g_{hj}\, \Gamma^h_{ik} - g_{ih}\, \Gamma^h_{jk} = 0, \tag{29.1}$$

where the Γ's are the Christoffel symbols of the second kind.

Conversely, if equations (29.1) for a given set of Γ's admit a solution $g_{ij}(i, j = 1, \cdots, n)$, then we have

$$\begin{aligned} & g_{ik,j} + g_{jk,i} - g_{ij,k} \\ &= \frac{\partial g_{ik}}{\partial x^j} + \frac{\partial g_{jk}}{\partial x^i} - \frac{\partial g_{ij}}{\partial x^k} - 2 g_{hk}\, \Gamma^h_{ij} = 0, \end{aligned} \tag{29.2}$$

from which it follows that the Γ's are the Christoffel symbols of the second kind formed with respect to the g's (cf. § 1). Consequently a necessary and sufficient condition that a space with an assigned symmetric connection be Riemannian is that equations (29.1) admit a solution.

From (6.4) we have that the conditions of integrability of equations (29.1) are

$$g_{ij,kl} - g_{ij,lk} = g_{ih} B^h_{jkl} + g_{hj} B^h_{ikl}, \tag{29.3}$$

which are reducible by means of (29.1) to

$$g_{ih} B^h_{jkl} + g_{hj} B^h_{ikl} = 0. \tag{29.4}$$

If these equations are not satisfied identically, that is, if we do not have $B^i_{jkl} = 0$, the solutions of these equations must satisfy (29.1). Differentiating (29.4) covariantly with respect to x^m and expressing the condition that (29.1) be satisfied, we have

$$g_{ih} B^h_{jkl,m} + g_{hj} B^h_{ikl,m} = 0. \tag{29.5}$$

Proceeding in this manner, we get the sequence of equations

$$\begin{aligned} g_{ih} B^h_{jkl,mm_1} + g_{hj} B^h_{ikl,mm_1} &= 0, \\ \cdots\cdots\cdots & \\ \cdots\cdots\cdots & \\ g_{ih} B^h_{jkl,mm_1\cdots m_r} + g_{hj} B^h_{ikl,mm_1\cdots m_r} &= 0, \\ \cdots\cdots\cdots & \end{aligned} \tag{29.6}$$

Because of the results of § 8 we have:

A necessary and sufficient condition that equations (29.1) *admit a solution is that there exist a positive integer N such that the first N sets of the equations* (29.4), (29.5) *and* (29.6) *admit a complete system of* r ($\geqq 1$) *sets of solutions which satisfy also the* $(N+1)$*th set; then the complete system can be chosen so that the functions* g_{ij} *of each set satisfy* (29.1).

Since equations (29.4), (29.5) and (29.6) are tensor equations, it follows that the numbers N and r, defined in the theorem, are invariant numbers for the connection.

If $r = 1$ and $\bar{g}_{ij}$ is a solution of the N sets of equations but not a solution of (29.1), then there exists a function φ such that the quantities $e^{-\varphi}\,\bar{g}_{ij}$ satisfy equations of the form (29.1). If the determinant $|\bar{g}_{ij}|$ is not zero, from equations (29.2) we have

(29.7) $$\Gamma^h_{ij} = \begin{Bmatrix} h \\ ij \end{Bmatrix} + \frac{1}{2}(\bar{g}_{ij}\,\bar{g}^{hk}\,\varphi_{,k} - \delta^h_i\,\varphi_{,j} - \delta^h_j\,\varphi_{,i}),$$

where $\bar{g}^{hk}$ is defined by

(29.8) $$\bar{g}^{hk}\,\bar{g}_{ik} = \delta^h_i,$$

and $\begin{Bmatrix} h \\ ij \end{Bmatrix}$ are Christoffel symbols of the second kind formed with respect to the $\bar{g}$'s. Consequently when the Γ's are expressible in the form (29.7), the space is Riemannian.* In terms of $\bar{g}_{ij}$, equations (29.1) become

(29.9) $$\bar{g}_{ij,k} = \bar{g}_{ij}\,\varphi_{,k}.$$

When $r>1$ and the solutions are $g^{(\alpha)}_{ij}$ for $\alpha = 1, \cdots, r$, then

(29.10) $$g_{ij} = A_{\alpha}\,g^{(\alpha)}_{ij},$$

where the A's are arbitrary constants, is a solution. In general the A's can be chosen so that the determinant $|g_{ij}|$ is not zero. When this solution g_{ij} is taken as the fundamental tensor of the Riemannian space, the tensor B^i_{jkl} becomes the Riemannian curvature tensor R^i_{jkl}. For this space the other $r-1$ sets of solutions are tensors whose first covariant derivatives are zero. Hence we have the theorem:

A necessary and sufficient condition that there exist for a Riemannian space p ($\geqq 1$) tensors $a^{(\alpha)}_{ij}$, other than the fundamental tensor g_{ij}, such that their first covariant derivatives be zero, is that there exist a positive integer N, such that the first N sets of equations (29.4), (29.5) *and* (29.6), *in which B^i_{jkl} is the Riemannian tensor R^i_{jkl}, admit a complete set of solutions, $g_{ij}, a^{(1)}_{ij}, \cdots, a^{(p)}_{ij}$, which also satisfy the* ($N+1$)*th set of the equations.*†

A space with a symmetric connection is said to be *flat*, or *plane*, if the curvature tensor B^i_{jkl} is zero. In this case equations (29.1) are completely integrable. Hence we have:

* Cf. *Eisenhart* and *Veblen*, 1922, 4, pp. 22, 23; also *Veblen* and *T. Y. Thomas*, 1923, 1, pp. 590, 591.

† Cf. *Eisenhart*, 1923, 3; also, *Levy*, 1926, 5.

A flat space is necessarily a Riemannian space.
From the last theorem of § 9 it follows that a preferred coördinate system exists for a flat space such that the coefficients Γ^i_{jk} for this coördinate system are everywhere zero. In this coördinate system the solutions g_{ij} of equations (29.1) are constants. Consequently the preferred coördinates are generalized cartesian coördinates.*

30. Symmetric connections of Weyl. We consider the symmetric connections for which there exists a vector φ_i and a symmetric tensor g_{ij} such that

$$g_{ij,k} + 2\, g_{ij}\, \varphi_k = 0, \tag{30.1}$$

and the determinant g is not zero. We remark that it follows from (29.9) that if φ_k is a gradient the space is Riemannian. We assume that φ_k is not a gradient.

If we substitute in (30.1) the expressions (5.9) for the Γ's, we obtain

$$g_{jh}\, a^h_{ik} + g_{ih}\, a^h_{jk} = 2 g_{ij}\, \varphi_k.$$

From these equations we obtain

$$a^i_{jk} = \delta^i_j\, \varphi_k + \delta^i_k\, \varphi_j - g_{jk}\, \varphi^i, \tag{30.2}$$

where

$$\varphi^i = g^{ij}\, \varphi_j. \tag{30.3}$$

Consequently the coefficients of the connection are

$$\Gamma^i_{jk} = \left\{ {i \atop jk} \right\} + \delta^i_j\, \varphi_k + \delta^i_k\, \varphi_j - g_{jk}\, \varphi^i. \tag{30.4}$$

Symmetric connections of this kind have been proposed by Weyl† as the basis of a combined theory of gravitation and electro-dynamics. From the remarks at the beginning of this section we observe that in a sense it is an immediate generalization of a Riemannian geometry.

If we put

$$\bar{g}_{ij} = e^{2\theta}\, g_{ij}, \qquad \bar{\varphi}_i = \varphi_i - \frac{\partial \theta}{\partial x^i}, \tag{30.5}$$

* Cf. 1926, 1, p. 84.

† 1921, 1, pp. 125, 296.

where θ is an arbitrary function of the x's, we have that $\bar{g}_{ij}$ and $\bar{\varphi}_i$ also furnish a solution of (30.1), when g_{ij} and φ_i do. We may speak of two Weyl geometries whose fundamental quantities are in the relations (30.5) as *conformal* to one another. Weyl refers to the effect of changing θ as a *change of gauge.*

When we express the conditions of integrability of equations (30.1), we have in consequence of (6.4)

$$(30.6) \qquad g_{hj}\, B^h_{ikl} + g_{ih}\, B^h_{jkl} + 2\, g_{ij}\, (\varphi_{k,l} - \varphi_{l,k}) = 0.$$

Multiplying by g^{ij} and summing for i and j, we have

$$(30.7) \qquad B^i_{ikl} = n\, (\varphi_{l,k} - \varphi_{k,l}) = n \left(\frac{\partial \varphi_l}{\partial x^k} - \frac{\partial \varphi_k}{\partial x^l} \right).$$

In § 5 it was seen that for any manifold B^i_{ikl} is the curl of a covariant vector which is determined to within an additive arbitrary gradient. In consequence of (30.7) we may consider φ_k in (30.1) as a definite function of the x's namely $\frac{1}{n} a_k$, where a_k is a vector whose curl is equal to B^i_{ilk} [cf. (5.10)]. Hence equations (30.1) are of the form (8.1).

By means of (30.7) equations (30.6) may be written in the form

$$(30.8) \qquad g_{hj}\, \bar{B}^h_{ikl} + g_{ih}\, \bar{B}^h_{jkl} = 0,$$

where

$$\bar{B}^h_{ikl} = B^h_{ikl} - \frac{1}{n}\, \delta^h_i\, B^j_{jkl}.$$

Equations (30.8) constitute the set F_1 for the theorem of § 8, and the sets $F_2, F_3, \cdots$ are obtained from (29.5) and (29.6) on replacing the B's by $\bar{B}$'s; we call them (30.8)′ and (30.8)″ respectively. Hence we have by means of § 8:

A necessary and sufficient condition that equations (30.1) *admit a solution is that φ_k be a vector such that B^i_{ikl} is the curl of the vector $n\varphi_l$ and that there exist a positive integer N such that the first N sets of equations* (30.8), (30.8)′ *and*

(30.8)″ *admit a complete system of* $r\ (\geqq 1)$ *sets of solutions which satisfy also the* $(N+1)$*th set; then the complete system can be chosen so that the functions* g_{ij} *of each system satisfy equations* (30.1).

When $r = 1$, we must add the further condition $|g_{ij}| \neq 0$, in order that a given connection be that of Weyl. When $r > 1$, ordinarily by a suitable choice of the constants A in equations of the form (29.10) we can obtain a solution g_{ij} for which $|g_{ij}| \neq 0$ and thus have a Weyl geometry. When more than one such solution exist, not conformal to one another, we have several geometries of Weyl, which have the same symmetric connection, and consequently the same paths.

As in § 29 we have that N and r are invariantive numbers of Weyl connections.

31. Homogeneous first integrals of the equations of the paths. If each integral of the equations of the paths (22.4) satisfies the condition

$$(31.1) \qquad a_{r_1 \cdots r_m} \frac{dx^{r_1}}{ds} \frac{dx^{r_2}}{ds} \cdots \frac{dx^{r_m}}{ds} = \text{const.},$$

the equations (22.4) are said to admit a homogeneous first integral of the mth degree. From the form of (31.1) it is seen that there is no loss of generality in assuming that the tensor $a_{r_1 \cdots r_m}$ is symmetric in all subscripts. If we differentiate (31.1) covariantly with respect to x^k, multiply by $\frac{dx^k}{ds}$, sum for k and make use of the equations of the paths written in the form

$$(31.2) \qquad \frac{dx^k}{ds}\left(\frac{dx^i}{ds}\right)_{,k} = 0,$$

we obtain

$$(31.3) \qquad a_{r_1 \cdots r_m, k} \frac{dx^{r_1}}{ds} \cdots \frac{dx^{r_m}}{ds} \frac{dx^k}{ds} = 0.$$

Since this equation must be satisfied identically (otherwise we should have all the solutions of (22.4) satisfying a differential equation of the first order), we must have

(31.4) $$P(a_{r_1 \cdots r_m, k}) = 0,$$

where P indicates the sum of $m+1$ terms obtained by permuting the subscripts cyclically. Conversely, if equations (31.4) are satisfied, equations (31.3) are and the left-hand member of (31.1) is constant along any path. (Cf. § 43).

In particular, if the integral is of the first degree, that is.

$$a_i \frac{dx^i}{ds} = \text{const.}.$$

the conditions (31.4) are

$$a_{i,j} + a_{j,i} = 0.$$

The question of linear first integrals is considered in § 44.

We consider the case when the equations of the paths admit a quadratic integral, namely

(31.5) $$g_{ij} \frac{dx^i}{ds} \frac{dx^j}{ds} = \text{const.}$$

In this case the conditions (31.4) are

(31.6) $$g_{ij,k} + g_{jk,i} + g_{ki,j} = 0.$$

From § 29 it is seen that Riemannian spaces are a sub-class of spaces with symmetric connection for which a homogeneous quadratic integral exists.

From (31.6) we have

(31.7) $$g_{ij,kl} + g_{jk,il} + g_{ki,jl} = 0.$$

Interchanging k and l, we have

(31.8) $$g_{ij,lk} + g_{jl,ik} + g_{li,jk} = 0.$$

If we subtract from the sum of these two equations the sum of the two equations obtained by interchanging i and k and j and k in (31.7), the resulting equation is reducible by means of (6.4), (21.3) and (21.4) to

(31.9) $$g_{ij,kl} - g_{kl,ij} = g_{\alpha j} B^{\alpha}_{lki} + g_{i\alpha} B^{\alpha}_{lkj} - g_{\alpha l} B^{\alpha}_{jik} - g_{k\alpha} B^{\alpha}_{jil}.$$

If k and l are interchanged and this equation is subtracted from (31.9), the resulting equation is satisfied because of (6.4). Thus we are unable to solve equations of the form (31.7) for each of the quantities $g_{ij,kl}$. However, if (31.7) be differentiated covariantly, we obtain equations which can be solved for $g_{ij,klm}$ and then the further conditions of integrability can be obtained with the aid of (6.4); and the determination of whether a given space admits one or more quadratic integrals is reducible to an algebraic problem somewhat after the manner of § 29, as has been shown by Veblen and T. Y. Thomas.* Instead of developing this question further we consider the problem of determining symmetric connections for which there is a given quadratic integral, such that the determinant g is not zero.

With the aid of the tensor g_{ij} we write the Γ's in the form (5.9); then

$$g_{ij,k} = -a_{kij} - a_{jki},$$

where

$$(31.10)\qquad a_{kij} = g_{jh}\, a^h_{ki}, \qquad a^i_{jk} = g^{ih}\, a_{jkh}, \qquad g^{ih}\, g_{hj} = \delta^i_j;$$

we remark that a^i_{jk} is symmetric in j and k, as follows from (5.9). Hence the conditions (31.6) become

$$(31.11)\qquad a_{ijk} + a_{jki} + a_{kij} = 0.$$

If c_{ijk} is any tensor symmetric in i and j and we put

$$(31.12)\qquad a_{ijk} = 2\,c_{ijk} - c_{ikj} - c_{jki},$$

the condition (31.11) is satisfied. Hence if we have any tensor c_{ijk} symmetric in i and j, and define a^i_{jk} by (31.10) and (31.12), then the symmetric connection given by (5.9) is such that the equations of the paths admit the first integral (31.5).

Conversely, if (31.5) is satisfied for a given connection and consequently a_{ijk} are given, the tensor c_{ijk} is not uniquely defined by (31.12). In fact, from (31.11) it follows that

* 1923, 1, pp. 599–608.

$a_{iii} = 0$ and consequently c_{iii} are arbitrary. When two of the indices are the same, we have from (31.11) and (31.12)

$$c_{iij} - c_{iji} = \frac{1}{2} a_{iij} = -a_{iji}.$$

Consequently either one of these c's can be chosen arbitrarily and the other is then determined; hence there are $n(n-1)$ arbitrary choices. When all of the indices are different and have given values, there are two independent equations (31.12) for the determinination of the c's with these same indices. Consequently any one may be taken arbitrarily and the others are uniquely determined. Hence we have:

A tensor g_{ij} for which $g \neq 0$ and a tensor c_{ijk}, symmetric in i and j determine a symmetric connection for which the equations of the paths admit the first integral (31.5); *conversely, if a geometry of paths is given whose equations admit a first integral, $n(n+1)(n+2)/6$ of the components c_{ijk} are arbitrary and the others are uniquely determined.**

* Cf. *Eisenhart*, 1924, 2, p. 384.

CHAPTER III

PROJECTIVE GEOMETRY OF PATHS

32. Projective change of affine connection. The Weyl tensor. In § 22 it was shown that the paths are the same for two symmetric connections whose coefficients are in the relations

$$\overline{\Gamma}^i_{jk} = \Gamma^i_{jk} + \delta^i_j \psi_k + \delta^i_k \psi_j, \tag{32.1}$$

where ψ_i is an arbitrary covariant vector. We say that the affine connection of coefficients $\overline{\Gamma}^i_{jk}$ is obtained from that with the coefficients Γ^i_{jk} by a *projective change* of the connection.

If we write the equations of the paths in the form

$$\frac{d^2 x^i}{d\overline{s}^2} + \overline{\Gamma}^i_{jk} \frac{dx^j}{d\overline{s}} \frac{dx^k}{d\overline{s}} = 0, \tag{32.2}$$

analogous to (22.4), we have from these equations and (32.1) that $\overline{s}$ is given as a function of s along any path by

$$\overline{s} = c \int e^{2\int \psi_k dx^k} ds. \tag{32.3}$$

For the expressions $\overline{\Gamma}^i_{jk}$ the components of the curvature tensor $\overline{B}^i_{jkl}$, analogous to (21.1), are reducible to

$$\overline{B}^i_{jkl} = B^i_{jkl} + \delta^i_j (\psi_{lk} - \psi_{kl}) + \delta^i_l \psi_{jk} - \delta^i_k \psi_{jl}, \tag{32.4}$$

where

$$\psi_{jk} = \psi_{j,k} - \psi_j \psi_k. \tag{32.5}$$

Contracting for i and l and for i and j, we have (cf. § 5)

$$\overline{B}_{jk} = B_{jk} + n\psi_{jk} - \psi_{kj}, \tag{32.6}$$

$$\begin{aligned} \overline{\beta}_{kl} &= \beta_{kl} + \frac{n+1}{2}(\psi_{kl} - \psi_{lk}) \\ &= \beta_{kl} + \frac{n+1}{2}\left(\frac{\partial \psi_k}{\partial x^l} - \frac{\partial \psi_l}{\partial x^k}\right). \end{aligned} \tag{32.7}$$

In § 5 it was shown that β_{ij} is the skew-symmetric part of B_{ij} and that it is the curl of a vector $a_i/2$. Consequently if we choose

$$\psi_k = -\frac{1}{n+1}\left(a_k + \frac{\partial \sigma}{\partial x^k}\right), \tag{32.8}$$

where σ is an arbitrary function of the x's, we have $\bar{\beta}_{kl} = 0$. Hence we have:

*By a suitable projective change of the affine connection the tensor B_{ij} for the new connection is symmetric and the tensor B^i_{ijk} is a zero tensor.**

From (32.7) we have also:

When the tensor B_{ij} is symmetric, a necessary and sufficient condition that the tensor $\bar{B}_{ij}$ for a projective change of connection be symmetric is that ψ_i be a gradient.

From equations (32.6) and (32.7) we have

$$\begin{aligned} \psi_{jk} &= \frac{1}{n-1}(\bar{B}_{jk} - B_{jk}) - \frac{2}{n^2-1}(\bar{\beta}_{jk} - \beta_{jk}), \\ \psi_{jk} - \psi_{kj} &= \frac{2}{n+1}(\bar{\beta}_{jk} - \beta_{jk}). \end{aligned} \tag{32.9}$$

When these expressions are substituted in (32.4), the latter are reducible to

$$\bar{W}^i_{jkl} = W^i_{jkl},$$

where

$$\begin{aligned} W^i_{jkl} = B^i_{jkl} &+ \frac{2}{n+1}\delta^i_j \beta_{kl} + \frac{1}{n-1}(\delta^i_k B_{jl} - \delta^i_l B_{jk}) \\ &+ \frac{2}{n^2-1}(\delta^i_l \beta_{jk} - \delta^i_k \beta_{jl}). \end{aligned} \tag{32.10}$$

Hence the tensor W^i_{jkl} is independent of the vector ψ_i, that is, it is unaltered by a projective change of affine connection. It was discovered by Weyl†, and was called by him the *projective curvature tensor*. We shall call it the *Weyl tensor*.

* *Eisenhart*, 1922, 2, p. 236.

† 1921, 2, p. 101.

Since

$$B_{ij} - B_{ji} = 2\beta_{ij}, \tag{32.11}$$

equations (32.10) can be written in the form

$$\begin{aligned} W^i_{jkl} &= B^i_{jkl} + \frac{1}{n+1}\delta^i_j(B_{kl} - B_{lk}) \\ &\quad + \frac{1}{n^2-1}[\delta^i_k(nB_{jl} + B_{lj}) - \delta^i_l(nB_{jk} + B_{kj})]. \end{aligned} \tag{32.12}$$

In consequence of the identities (21.3) and (21.4) we have the identities

$$W^i_{jkl} + W^i_{jlk} = 0, \tag{32.13}$$

$$W^i_{jkl} + W^i_{klj} + W^i_{ljk} = 0. \tag{32.14}$$

Also from (32.10) we have by contraction

$$W^i_{ikl} = W^i_{jki} = 0. \tag{32.15}$$

If we differentiate (32.10) covariantly with respect to the Γ's and make use of the identities (21.5), we obtain

$$\begin{aligned} &W^i_{jkl,m} + W^i_{jlm,k} + W^i_{jmk,l} \\ &\quad = \frac{1}{n-1}[\delta^i_k(B_{jl,m} - B_{jm,l}) + \delta^i_l(B_{jm,k} - B_{jk,m}) \\ &\qquad\qquad + \delta^i_m(B_{jk,l} - B_{jl,k})] \\ &\quad + \frac{2}{n^2-1}[\delta^i_k(\beta_{jm,l} - \beta_{jl,m}) + \delta^i_l(\beta_{jk,m} - \beta_{jm,k}) \\ &\qquad\qquad + \delta^i_m(\beta_{jl,k} - \beta_{jk,l})] \end{aligned}$$

Contracting for i and m, we have in consequence of (32.15),

$$W^i_{jkl,i} = \frac{n-2}{n-1}\left[B_{jk,l} - B_{jl,k} + \frac{2}{n+1}(\beta_{jl,k} - \beta_{jk,l})\right] \tag{32.16}$$

An invariant, such as the Weyl tensor, which is unaltered by any projective change of the affine connection is called a *projective invariant*. By processes analogous to those used in § 18 we establish the theorem:

If $\lambda^i_{(\alpha)}$ and $\lambda^{(\alpha)}_i$ are the components of any ennuple, the tensors

$$\lambda^i_{(\alpha),j} - \frac{1}{n+1}(\lambda^i_{(\alpha)}\lambda^{(\beta)}_k \lambda^k_{(\beta),j} + \delta^i_j \lambda^h_{(\alpha)}\lambda^{(\beta)}_k \lambda^k_{(\beta),h})$$

and the invariants

$$\gamma_{\mu}{}^{\nu}{}_{\sigma} - \frac{1}{n+1}(\delta^{\nu}_{\mu}\gamma_{\alpha}{}^{\alpha}{}_{\sigma} + \delta^{\nu}_{\sigma}\gamma_{\alpha}{}^{\alpha}{}_{\mu})$$

*are projective invariants.**

A tensor which is not a projective invariant may, however, be invariant under certain projective changes of the connection. Thus from (32.4) and (32.5) we derive the theorem:

A necessary and sufficient condition that the curvature tensor be unaltered by the projective change defined by a vector ψ_i is that the latter satisfy the condition

$$\psi_{i,j} - \psi_i \psi_j = 0. \tag{32.17}$$

This is also a necessary and sufficient condition that the tensor B_{ij} is invariant under the change.†

Equations (32.17) can be written in the form

$$\frac{\partial \psi_i}{\partial x^j} - \psi_h \hat{\Gamma}^h_{ij} = 0,$$

where

$$\hat{\Gamma}^h_{ij} = \Gamma^h_{ij} + \frac{1}{2}(\delta^h_i \psi_j + \delta^h_j \psi_i). \tag{32.18}$$

Hence the space with the affine connection defined by $\hat{\Gamma}^h_{ij}$ admits a field of parallel covariant vectors ψ_i (§ 11). Consequently the problem of the theorem and that of spaces admitting fields of parallel covariant vectors are equivalent.

Similarly from (32.5) and (32.6) we have:

A necessary and sufficient condition that the symmetric part of the tensor B_{ij} be unaltered by the projective change defined by a vector ψ_i is that

$$\psi_{i,j} + \psi_{j,i} - 2\psi_i \psi_j = 0.\ddagger \tag{32.19}$$

* Cf. *T. Y. Thomas*, 1925, 10, p. 319; also *Levy*, 1927, 1, p. 310.

† Cf. *Schouten*, 1925, 6, p. 453; also, *J. M. Thomas*, 1926, 8, p. 62.

‡ Cf. *J. M. Thomas*, 1926, 8, p. 62.

For the affine connection defined by (32.18) equations (32.19) become

$$\psi_{i,\hat{j}} + \psi_{j,\hat{i}} = 0$$

that is, the equations of the paths admit the linear first integral $\psi_i \frac{dx^i}{d\hat{s}} = \text{const.}$, where $\hat{s}$ is defined by an equation of the form (32.3) (cf. § 43).

Also we have from (32.7):

A necessary and sufficient condition that the skew-symmetric part of the tensor B_{ij}, and consequently the tensor B^h_{hij}, be unaltered by the projective change defined by a vector ψ_i is that ψ_i be the gradient of an arbitrary function.

The second theorem in this section is a corollary of the above theorem.

33. Affine normal coördinates under a projective change of connection. If we denote by y^i and y'^i the affine normal coördinates corresponding to the same coördinate system x^i for two connections in the relation (32.1), the equations of the paths through the origin P in these coördinate systems are given by (23.1) and

$$y'^i = \left(\frac{dx^i}{d\bar{s}}\right)_0 \bar{s}.$$

Since a projective change of connection leaves each path individually invariant, it follows from the above equations of the paths that along each path y'^i is proportional to y^i. Moreover, throughout the domain under consideration y'^i is a function of the y's. Consequently these functions must be of the form

$$y'^i = \frac{y^i}{f(y)}, \tag{33.1}$$

where $f(y)$ is a function of the y's regular in the neighborhood of the origin and not vanishing at the origin. Similarly we have

$$y^i = \frac{y'^i}{f'(y')}; \tag{33.2}$$

where f' is of similar character and $f(y)\cdot f'(y')=1$, because of (33.1) and (33.2).

If C^i_{jk} and $\bar{C}^i_{jk}$ are the respective coefficients of connection in the y's, we have

$$\bar{C}^i_{jk} = C^i_{jk} + \delta^i_j \psi_k + \delta^i_k \psi_j, \tag{33.3}$$

where ψ_j are the components in the y's of the vector defining the projective change. If $\bar{C}'^i_{jk}$ are the coefficients in the y''s of the second connection, we have from equations of the form (5.6)

$$\bar{C}'^\alpha_{\beta\gamma} = \left(\frac{\partial^2 y^i}{\partial y'^\beta \partial y'^\gamma} + \bar{C}^i_{jk}\frac{\partial y^j}{\partial y'^\beta}\frac{\partial y^k}{\partial y'^\gamma}\right)\frac{\partial y'^\alpha}{\partial y^i} \tag{33.4}$$

Since the y''s are normal coördinates, we have $\bar{C}'^\alpha_{\beta\gamma} y'^\beta y'^\gamma = 0$ (§ 23). In consequence of (33.4) these become

$$\left(\bar{C}^i_{jk}\frac{\partial y'^\alpha}{\partial y^i} - \frac{\partial^2 y'^\alpha}{\partial y^j \partial y^k}\right)\frac{\partial y^j}{\partial y'^\beta}\frac{\partial y^k}{\partial y'^\gamma} y'^\beta y'^\gamma = 0. \tag{33.5}$$

From (33.2) we have

$$\frac{\partial y^j}{\partial y'^\beta} = \frac{1}{f'^2}\left(\delta^j_\beta f' - y'^j \frac{\partial f'}{\partial y'^\beta}\right),$$

$$\frac{\partial y^j}{\partial y'^\beta} y'^\beta = \frac{y'^j}{f'^2}\left(f' - \frac{\partial f'}{\partial y'^\beta} y'^\beta\right),$$

and from (33.1)

$$\frac{\partial y'^\alpha}{\partial y^i} = \frac{1}{f^2}\left(\delta^\alpha_i f - y^\alpha \frac{\partial f}{\partial y^i}\right),$$

$$\frac{\partial^2 y'^\alpha}{\partial y^j \partial y^k} y^j y^k = -\frac{y^\alpha}{f^2}\left[2 y^j \frac{\partial f}{\partial y^j} + \frac{\partial^2 f}{\partial y^j \partial y^k} y^j y^k - \frac{2}{f}\left(\frac{\partial f}{\partial y^k} y^k\right)^2\right].$$

Since the equation $f' - \dfrac{\partial f'}{\partial y'^\beta} y'^\beta = 0$ does not admit a solution regular at the origin and not vanishing there, and since $C^i_{jk} y^j y^k = 0$, the equations (33.5) are reducible to

$$\left(y^k \psi_k + \frac{1}{f}\frac{\partial f}{\partial y^j} y^j\right)\left(f - \frac{\partial f}{\partial y^i} y^i\right) + \frac{1}{2}\frac{\partial^2 f}{\partial y^i \partial y^j} y^i y^j = 0. \tag{33.6}$$

Since ψ_k are assumed to be regular at P, we put

$$\psi_k = b_{k0} + b_{ki}\, y^i + \frac{1}{2!} b_{kij}\, y^i\, y^j + \cdots,$$

$$f = 1 + a_i\, y^i + \frac{1}{2!} a_{ij}\, y^i\, y^j + \cdots,$$

where without loss of generality the a's are symmetric in the indices and the b's in all but the first. Substituting in (33.6), we find that $a_i = -b_{i0}$ and that the other a's are uniquely determined. Thus when ψ_k are given, the function f is determined.

There is also the converse problem of giving f and finding the ψ's from (33.6). We consider in particular the case when

$$f = 1 + a_i\, y^i, \tag{33.7}$$

where the a's are constants, so that the transformation (33.1) is linear fractional. Now equation (33.6) reduces to

$$y^k\, \psi_k + \frac{a_j\, y^j}{1 + a_i\, y^i} = 0. \tag{33.8}$$

Although this equation gives the condition on the ψ's in the y's, we are interested in finding their components in a general coördinate system so that we may have a means to knowing when the case (33.7) is possible. To this end we differentiate this equation with respect to y^l, multiply by y^l and sum for l; then from the resulting equation and (33.8) we eliminate $a_i\, y^i$. This gives

$$y^k\, y^l \left(\frac{\partial \psi_k}{\partial y^l} - \psi_k\, \psi_l \right) = 0, \tag{33.9}$$

which because of the relation $C^i_{jk}\, y^j\, y^k = 0$ and (23.1) can be written thus

$$\frac{dy^k}{ds} \frac{dy^l}{ds} (\psi_{k,l} - \psi_k\, \psi_l) = 0.$$

In the general coördinate system x^i corresponding to the y's this equation is

$$\frac{d x^k}{ds} \frac{d x^l}{ds} (\psi'_{k,l} - \psi'_k \psi'_l) = 0,$$

where ψ'_k are the components in the x's of the vector ψ_k in the y's. If this condition is to be satisfied at each point in space, it is necessary that

$$\psi'_{k,l} + \psi'_{l,k} - 2\psi'_k \psi'_l = 0. \tag{33.10}$$

Since this is a tensor equation, if it holds in one coördinate system, it holds in all. Conversely, if there exists a vector satisfying (33.10) for a given space, and we choose a normal coördinate system y^i with a given point P for origin, equation (33.9) holds at P. If we put

$$y^k \psi_k + \frac{f-1}{f} = 0, \tag{33.11}$$

differentiate with respect to y^l, multiply by y^l and sum for l, we have in consequence of (33.9)

$$y^k \psi_k (y^l \psi_l + 1) + \frac{1}{f^2} \frac{\partial f}{\partial y^l} y^l = 0.$$

By means of (33.11) this is reducible to

$$f - \frac{\partial f}{\partial y^l} y^l = 1. \tag{33.12}$$

If the function f is regular at P, then $f = 1$ at P and the integral of (33.12) satisfying this condition is $f = 1 + a_i y^i$.

From these results and the fifth theorem of § 32 we have:

*In order that the affine normal coördinates at every point undergo a linear fractional transformation when the affine connection undergoes a projective change, it is necessary and sufficient that the symmetric part of the tensor B_{ij} be unaltered by the projective change.**

34. Projectively flat spaces. We may interpret the results of § 32 as giving spaces with corresponding paths.

* The question of this type of projective change was raised by *Veblen*, 1925, 7, p. 131, and the theorem was established, in a different manner, by *J. M. Thomas*, 1926, 8, p. 62.

Weyl* has called a space V_n *projectively flat* when its paths have the same equations as the paths of a flat space $\overline{V}_n$ (§ 29). This is equivalent to saying that for V_n there exists a preferred coördinate system in terms of which the finite equations of the paths are linear.

Since for $\overline{V}_n$ we have

$$(34.1) \qquad \overline{B}^i_{jkl} = 0, \quad \overline{B}_{ij} = 0, \quad \overline{\beta}_{ij} = 0,$$

it is evident that a necessary condition that a space be projectively flat is that the Weyl tensor be zero. We shall show that this condition is also sufficient, when $n > 2$.

From the first of equations (32.9) we have

$$(34.2) \qquad \psi_{i,j} = \psi_i \psi_j - \frac{1}{n-1} B_{ij} + \frac{2}{n^2-1} \beta_{ij}.$$

The conditions of integrability of these equations, namely

$$\psi_{i,jk} - \psi_{i,kj} = \psi_h B^h_{ijk},$$

are reducible, by means of (32.11) and the expression for B^i_{jkl} obtained by equating to zero the right-hand member of (32.10), to

$$(34.3) \qquad B_{ik,j} - B_{ij,k} + \frac{2}{n+1} (\beta_{ij,k} - \beta_{ik,j}) = 0.$$

From (32.16) it follows that these equations are a consequence of the vanishing of the Weyl tensor, when $n > 2$, as was to be proved.

When $n = 2$, we have, because of the identities (21.3),

$$B_{11} = B^2_{112}, \quad B_{12} = B^1_{121}, \quad B_{21} = B^2_{212}, \quad B_{22} = B^1_{221}.$$

Hence from (32.12) we find:

The Weyl tensor vanishes identically when $n = 2$.

Accordingly we have the following theorem of Weyl:†

* 1921, 2, p. 104.

† 1921, 2, p. 105.

A necessary and sufficient condition that a space V_n with an affine connection be projectively flat is that the Weyl tensor vanish when $n > 2$ and that equations (34.3) be satisfied when $n = 2$.

From (32.1) it follows that, if a space V_n is projectively flat and x^i are cartesian coördinates in the projectively related flat space, the coefficients of the affine connection in V_n are given by

$$\Gamma^i_{jk} = -(\delta^i_j \psi_k + \delta^i_k \psi_j). \tag{34.4}$$

Conversely, the most general projectively flat space is obtained by taking the Γ's in the form (34.4), where ψ_j is an arbitrary vector.

When the expressions (34.4) are substituted in the equations of the paths (22.4), the latter can be integrated in the form

$$\frac{x^1 - a^1}{b^1} = \cdots = \frac{x^n - a^n}{b^n} = \int e^{2\int \psi_k dx^k} ds, \tag{34.5}$$

the integral $\int \psi_k \, dx^k$ being taken along a path, which result is in keeping with the remark at the beginning of the section.

From (34.2) it follows that a necessary and sufficient condition that V_n be a projectively flat space for which the tensor B_{ij} is symmetric is that ψ_j in (34.4) be a gradient. If we replace ψ_j in (34.4) by $\frac{\partial \psi}{\partial x^j}$, the components of the curvature tensor are expressible in the form

$$B^h_{ijk} = e^{-\psi} \left(\delta^h_j \frac{\partial^2 e^\psi}{\partial x^i \partial x^k} - \delta^h_k \frac{\partial^2 e^\psi}{\partial x^i \partial x^j} \right) \tag{34.6}$$

Contracting for h and k, we have

$$B_{ij} = (1 - n) e^{-\psi} \frac{\partial^2 e^\psi}{\partial x^i \partial x^j}. \tag{34.7}$$

From these equations we have

$$B^h_{ijk} + \frac{1}{n-1} (\delta^h_j B_{ik} - \delta^h_k B_{ij}) = 0. \tag{34.8}$$

The left-hand member of this equation is the expression for W^h_{ijk} when B_{ij} is symmetric, as follows from (32.10).

We consider now the question of determining the Riemannian projectively flat spaces, that is, spaces whose geodesics can be put into correspondence with the straight lines of a flat space. In this case B_{ij} are symmetric and are in fact R_{ij}, the components of the Ricci tensor.* For $n > 2$ we have equations (34.8), which are equivalent to

$$(34.9) \qquad R_{hijk} = \frac{1}{1-n}(g_{hj} R_{ik} - g_{hk} R_{ij}).$$

When in (34.9) we put $h = i$, we find that

$$(34.10) \qquad R_{ij} = K_0 (1-n) g_{ij},$$

where K_0 is the factor of proportionality thus obtained. By reason of (34.10) equations (34.9) are reducible to

$$(34.11) \qquad R_{hijk} = K_0 (g_{hj} g_{ik} - g_{hk} g_{ij}).$$

Consequently V_n for $n > 2$ is a space of constant Riemannian curvature K_0.†

When $n = 2$, it follows from the definition of R_{ij} that

$$\frac{R_{11}}{g_{11}} = \frac{R_{12}}{g_{12}} = \frac{R_{22}}{g_{22}} = \frac{R_{2112}}{g}.$$

Since R_{2112}/g is the Gaussian curvature K_0 of the surface, it follows that (34.10) holds also when $n = 2$. When we apply the conditions (34.3) to (34.10), we find that K_0 is a constant. Hence we have:

A necessary and sufficient condition that a Riemannian space be projectively flat is that its Riemannian curvature be constant.‡

From (34.10) we have for all values of n, $R_{ij,k} = 0$. Conversely, if we have

* 1926, 1, p. 21.

† 1926, 1, p. 83.

‡ Weyl, 1921, 2, p. 110.

$$(34.12) \qquad B_{ij} = B_{ji}, \quad B_{ij,k} = 0,$$

g_{ij} defined by (34.10), where K_0 is an arbitrary constant is such that $g_{ij,k} = 0$. Hence we have:

A necesssary and sufficient condition that a projectively flat space be Riemannian is that (34.12) *be satisfied.*

In the coördinate system for which R_{ij} takes the form (34.7) we have from (34.10)

$$(34.13) \qquad g_{ij} = \frac{1}{K_0} e^{-\psi} \frac{\partial^2 e^{\psi}}{\partial x^i \partial x^j}.$$

From this expression and (34.4) in which $\psi_j = \frac{\partial \psi}{\partial x^j}$ it follows that the conditions $g_{ij,k} = 0$ are reducible to

$$\frac{\partial^3 e^{2\psi}}{\partial x^i \partial x^j \partial x^k} = 0.$$

Consequently

$$(34.14) \qquad e^{2\psi} = a_{ij} x^i x^j + 2 b_i x^i + c,$$

where the a's, b's and c are constants.

35. Coefficients of a projective connection. From equations (32.1) we have

$$(35.1) \qquad \bar{\Gamma}^i_{ik} = \Gamma^i_{ik} + (n+1)\psi_k,$$

from which and (32.1) we find that the quantities

$$(35.2) \qquad \Pi^i_{jk} = \Gamma^i_{jk} - \frac{1}{n+1}(\delta^i_j \Gamma^h_{hk} + \delta^i_k \Gamma^h_{hj})$$

are independent of a projective change of affine connection.* We call Π^i_{jk} the *coefficients of a projective connection.*

In order to find the relations between the functions Π^i_{jk} and the analogous functions Π'^i_{jk} in a coördinate system x'^i, we remark that from equations of the form (5.6) we have

$$(35.3) \qquad \Gamma^i_{ij} = \Gamma'^{\alpha}_{\alpha\beta} \frac{\partial x'^{\beta}}{\partial x^j} + \frac{\partial \log \Delta}{\partial x^j},$$

* Cf. *T. Y. Thomas,* 1925, 2, p. 200.

where Δ is the jacobian of the transformation, that is,

$$\Delta = \left| \frac{\partial x'^i}{\partial x^j} \right|. \tag{35.4}$$

Then from (35.2) and analogous expression for $\Pi'^{\alpha}_{\beta\gamma}$ and from (5.6) and (35.3) we obtain

$$\begin{aligned} \frac{\partial^2 x'^{\alpha}}{\partial x^i \partial x^j} = \Pi^h_{ij} \frac{\partial x'^{\alpha}}{\partial x^h} - \Pi'^{\alpha}_{\beta\gamma} \frac{\partial x'^{\beta}}{\partial x^i} \frac{\partial x'^{\gamma}}{\partial x^j} &+ \frac{\partial x'^{\alpha}}{\partial x^j} \frac{\partial \theta}{\partial x^i} \\ &+ \frac{\partial x'^{\alpha}}{\partial x^i} \frac{\partial \theta}{\partial x^j}, \end{aligned} \tag{35.5}$$

where for the sake of brevity we have put

$$\theta = \frac{1}{n+1} \log \Delta. \tag{35.6}$$

When we express the conditions of integrability of equations (35.5), we obtain

$$\Pi^h_{ijk} = \Pi'^{\alpha}_{\beta\gamma\delta} \frac{\partial x^h}{\partial x'^{\alpha}} \frac{\partial x'^{\beta}}{\partial x^i} \frac{\partial x'^{\gamma}}{\partial x^j} \frac{\partial x'^{\delta}}{\partial x^k} - \delta^h_j c_{ik} + \delta^h_k c_{ij}, \tag{35.7}$$

where

$$\Pi^h_{ijk} = \frac{\partial \Pi^h_{ik}}{\partial x^j} - \frac{\partial \Pi^h_{ij}}{\partial x^k} + \Pi^l_{ik} \Pi^h_{lj} - \Pi^l_{ij} \Pi^h_{lk} \tag{35.8}$$

and

$$c_{ij} = \Pi^h_{ij} \frac{\partial \theta}{\partial x^h} + \frac{\partial \theta}{\partial x^i} \frac{\partial \theta}{\partial x^j} - \frac{\partial^2 \theta}{\partial x^i \partial x^j}. \tag{35.9}$$

Contracting (35.7) for h and k, we have

$$c_{ij} = \frac{1}{n-1} \left(\Pi_{ij} - \Pi'_{\beta\gamma} \frac{\partial x'^{\beta}}{\partial x^i} \frac{\partial x'^{\gamma}}{\partial x^j} \right), \tag{35.10}$$

where by definition

$$\Pi_{ij} = \Pi^h_{ijh}. \tag{35.11}$$

When the expressions (35.10) are substituted in (35.7), we have

$$W^h_{ijk} = W'^{\alpha}_{\beta\gamma\delta} \frac{\partial x^h}{\partial x'^{\alpha}} \frac{\partial x'^{\beta}}{\partial x^i} \frac{\partial x'^{\gamma}}{\partial x^j} \frac{\partial x'^{\delta}}{\partial x^k},$$

where

$$(35.12)\qquad W^h_{ijk} = \Pi^h_{ijk} + \frac{1}{n-1}(\delta^h_j \Pi_{ik} - \delta^h_k \Pi_{ij}).$$

From (35.8) and (35.2) we have

$$(35.13)\quad \Pi^h_{ijk} = B^h_{ijk} + \frac{1}{n+1}(2\,\delta^h_i \beta_{jk} - \delta^h_k A_{ij} + \delta^h_j A_{ik}),$$

where by definition

$$A_{ij} = \frac{\partial \Gamma^r_{ri}}{\partial x^j} - \Gamma^h_{ij}\Gamma^r_{rh} + \frac{1}{n+1}\Gamma^r_{ri}\Gamma^h_{hj},$$

and from (35.13)

$$(35.14)\quad \Pi_{ij} = B_{ij} - \frac{1}{n+1}[2\beta_{ij} + (n-1)A_{ij}].$$

When these expressions are substituted in (35.12), it is found that W^h_{ijk} so defined are the components of the Weyl tensor (32.10).*

Substituting the expression (35.10) for c_{ij} in (35.9), we obtain

$$(35.15)\ \frac{\partial^2 e^{-\theta}}{\partial x^i \partial x^j} = \Pi^l_{ij}\frac{\partial e^{-\theta}}{\partial x^l} + \frac{e^{-\theta}}{n-1}\left(\Pi_{ij} - \Pi'_{\beta\gamma}\frac{\partial x'^\beta}{\partial x^i}\frac{\partial x'^\gamma}{\partial x^j}\right).$$

Expressing the condition of integrability of these equations, we obtain

$$(35.16)\quad \Pi_{ijk} = \Pi'_{\alpha\beta\gamma}\frac{\partial x'^\alpha}{\partial x^i}\frac{\partial x'^\beta}{\partial x^j}\frac{\partial x'^\gamma}{\partial x^k} + (1-n)\,W^l_{ijk}\frac{\partial \theta}{\partial x^l},$$

where

$$(35.17)\quad \Pi_{ijk} = \frac{\partial \Pi_{ij}}{\partial x^k} - \frac{\partial \Pi_{ik}}{\partial x^j} + \Pi^l_{ij}\Pi_{lk} - \Pi^l_{ik}\Pi_{lj}.$$

36. The equivalence of projective connections. If we put

$$(36.1)\qquad \frac{\partial x'^\alpha}{\partial x^i} = u^\alpha_i, \qquad \frac{\partial \theta}{\partial x^i} = \varphi_i,$$

* Cf. *J. M. Thomas*, 1925, 3, p. 208.

equations (35.5) and (35.15) become

$$(36.2)\quad \begin{aligned} \frac{\partial u_i^\alpha}{\partial x^j} &= \Pi_{ij}^h u_h^\alpha - \Pi'^\alpha_{\beta\gamma} u_i^\beta u_j^\gamma + u_j^\alpha \varphi_i + u_i^\alpha \varphi_j, \\ \frac{\partial \varphi_i}{\partial x^j} &= \varphi_i \varphi_j + \Pi_{ij}^l \varphi_l + \frac{1}{1-n}(\Pi_{ij} - \Pi'_{\beta\gamma} u_i^\beta u_j^\gamma). \end{aligned}$$

These equations together with the functional relation (35.6) must admit a solution for given expressions of Π_{jk}^i and Π'^i_{jk} in the x's and x''s respectively, if the two projective connections so defined are to be equivalent. From the preceding results it follows that the conditions of integrability of these equations are

$$(36.3)\quad W_{ijk}^h u_h^\alpha = W'^\alpha_{\beta\gamma\delta} u_i^\beta u_j^\gamma u_k^\delta$$

$$(36.4)\quad \Pi_{ijk} = \Pi'_{\alpha\beta\gamma} u_i^\alpha u_j^\beta u_k^\gamma + (1-n) W_{ijk}^l \varphi_l.$$

If we denote by $\bar{u}_i^h$ the cofactor of u_h^i in the jacobian $|u_h^i|$ divided by the jacobian, we have from (35.6)

$$\frac{\partial \theta}{\partial x^j} = \frac{1}{n+1} \bar{u}_\alpha^i \frac{\partial u_i^\alpha}{\partial x^j},$$

which is satisfied identically because of (36.1) and (36.2).

If we differentiate equations (36.3) with respect to x^l, the resulting equations are reducible by means of (36.2) and (36.3) to

$$(36.5)\quad \begin{aligned} &u_h^\alpha (W_{ijk|l}^h - 2 W_{ijk}^h \varphi_l - W_{ljk}^h \varphi_i - W_{ilk}^h \varphi_j - W_{ijl}^h \varphi_k) \\ &\qquad + u_l^\alpha W_{ijk}^h \varphi_h = W'^\alpha_{\beta\gamma\delta|\varrho} u_i^\beta u_j^\gamma u_k^\delta u_l^\varrho, \end{aligned}$$

where $W_{ijk|l}^h$ denotes the *projective derivative* of W_{ijk}^h, that is, the covariant derivative with respect to the Π_{jk}^i; we remark that the projective derivative of a tensor is not in general a tensor. In this notation equations (35.17) may be written in the form

$$(36.6)\quad \Pi_{ijk} = \Pi_{ij|k} - \Pi_{ik|j}.$$

We observe from (35.12) and (32.10) that W_{ijk}^h is of the same form in Π_{jk}^i as it is in Γ_{jk}^i, when the tensor B_{ij} is symmetric. Since (32.16) follows formally from (32.10), we have from (35.12) and (36.6)

$$W^l_{ijk|l} = \frac{n-2}{n-1} \Pi_{ijk}. \tag{36.7}$$

If we multiply (36.5) by $\bar{u}^l_\alpha$ and sum for α and l, we obtain

$$W'^l_{ijk|l} + (n-2)\, W^l_{ijk}\, \varphi_l = W'^\varrho_{\beta\gamma\delta|\varrho}\, u^\beta_i\, u^\gamma_j\, u^\delta_k.$$

When $n > 2$, these equations are reducible by means of (36.7) to (36.4). Consequently in applying the results of § 8 we denote by F_1 equations (36.3) and proceed with these equations to get the sequence $F_2, F_3, \cdots$ of derived sets. Hence we have:

A necessary and sufficient condition that two projective connections for $n > 2$ be equivalent is that there exist a positive integer N, such that equations (35.6) *and the sets of equations $F_1, \cdots, F_N$ are compatible in θ, $\overset{\alpha}{u_i}$, x'^i and φ_i as functions of the x's, and that the $(N+1)$th set is satisfied in consequence of the preceding ones.**

When $n = 2$, W^h_{ijk} vanishes identically (§ 34). The above theorem applies to this case with the understanding that the sets $F_1, \cdots, F_N$ consist of (36.4) and the derived equations.

As in § 28 it can be shown that N is an invariantive number for all manifolds with the same projective connection, and likewise p, where $(n+1)^2 - p$ is the number of independent equations in the sets $F_0, \cdots, F_N$.

When $n > 2$, the Weyl tensor and its first, $\cdots$, Nth ($N \overline{\lessgtr} (n+1)^2$) projective derivatives form a complete system of projective invariants (§ 28) for the manifold. When $n = 2$, the functions Π_{ijk} and their first, $\cdots$, Nth ($N \overline{\lessgtr} 9$) projective derivatives form a complete system.

Another interpretation can be given to the preceding results. Thus let Π^i_{jk} be the coefficients of projective connection of a manifold V_n in coördinates x^i and similarly Π'^i_{jk} in any other coördinate system x'^i, the equations of the transformation being

$$x'^i = \varphi^i(x^1, \cdots, x^n). \tag{36.8}$$

* Cf. *Veblen* and *J. M. Thomas*, 1926, 6, pp. 288, 290.

We consider an associated manifold $^*V_{n+1}$, of coördinates $x^0, x^1, \cdots, x^n$ and in the x's define a set of functions $^*\Gamma^\alpha_{\beta\gamma}$ in $^*V_{n+1}$ by

$$(36.9)\quad ^*\Gamma^i_{jk} = \Pi^i_{jk}, \quad ^*\Gamma^\alpha_{0\beta} = -\frac{1}{n+1}\delta^\alpha_\beta, \quad ^*\Gamma^0_{ij} = \frac{1+n}{1-n}\Pi_{ij},$$

where greek indices take the values $0, 1, \cdots, n$ and latin $1, \cdots, n$. For the transformation in $^*V_{n+1}$ defined by (36.8) and

$$(36.10)\qquad x'^0 = x^0 + \log \Delta,$$

where Δ is given by (35.4), we find that the coefficients (36.9) and similar expressions in the x''s satisfy the relations

$$(36.11)\quad ^*\Gamma^\alpha_{\beta\gamma} = \frac{\partial x^\alpha}{\partial x'^\varrho}\left(\frac{\partial^2 x'^\varrho}{\partial x^\beta\, \partial x^\gamma} + {}^*\Gamma'^\varrho_{\sigma\tau}\frac{\partial x'^\sigma}{\partial x^\beta}\frac{\partial x'^\tau}{\partial x^\gamma}\right).$$

In fact, when α, β, γ take the values $1, \cdots, n$, equations (36.11) reduce to (35.5); when β or $\gamma = 0$, the equations are satisfied identically; and when $\alpha = 0$, and β and γ take values 1 to n, the equations reduce to (35.15). Thus the problem of equivalence of projective connections is reducible to a restricted problem for affine connections, as shown by T. Y. Thomas.†

In order to consider the problem more fully from this point of view, we denote by $^*B^\alpha_{\beta\gamma\delta}$ the curvature tensor formed with respect to the $^*\Gamma$'s. In consequence of (35.12) and (35.17) we have

$$^*B^i_{jkl} = W^i_{jkl}, \quad ^*B^0_{jlk} = \frac{n+1}{n-1}\Pi_{jkl}$$

and that all the other $^*B$'s vanish identically. If we take the functions so defined and apply the reasoning of § 28, equations of the form (28.2) (interchanging x's and x''s) become

$$(36.12)\quad W^h_{jkl}\, u^i_h + \frac{n+1}{n-1}\Pi_{jkl}\, u^i_0 = W'^i_{pqr}\, u^p_j\, u^q_k\, u^r_l,$$

† 1926, 10.

$$(36.13)\qquad \Pi_{jkl}\, u_0^0 + \frac{n-1}{n+1}\, u_h^0 W_{jkl}^h = \Pi'_{pqr}\, u_j^p\, u_k^q\, u_l^r,$$

$$(36.14)\qquad W'^h_{jkl}\, u_\varrho^j\, u_\sigma^k\, u_\tau^l = 0, \qquad \Pi'_{jkl}\, u_\varrho^j\, u_\sigma^k\, u_\tau^l = 0,$$

where one or more of the indices ϱ, σ, τ is 0.

For the relations (36.8) and (36.10) between the coördinates, equations (36.12) and (36.13) reduce to (36.3) and (36.4) respectively, and (36.14) are satisfied identically.

37. Normal affine connection. If for a given affine connection and a given coördinate system we choose for the components ψ_i of a projective change the values $-\Gamma_{hi}^h/(n+1)$, it follows from (35.1) that the coefficients of the new connection satisfy the conditions $\bar{\Gamma}_{hi}^h = 0$. We call this uniquely determined connection the *normal affine* connection for the given coördinate system.* Hence we have:

Among all the affine connections with the same projective connection there is a unique normal affine connection for any coördinate system.

From (5.8) it follows that B_{ij} is symmetric for a normal connection. Conversely, if B_{ij} is symmetric for an affine connection, we have from (5.8) in any coördinate system

$$(37.1)\qquad \frac{\partial \Gamma_{hj}^h}{\partial x^k} - \frac{\partial \Gamma_{hk}^h}{\partial x^j} = 0.$$

If we put

$$\frac{\partial \varphi}{\partial x^i} = \Gamma_{hi}^h,$$

these equations are completely integrable in consequence of (37.1). When we define a coördinate system x'^i by the equations

$$(37.2)\qquad x'^1 = \int e^\varphi\, dx^1, \quad x'^\alpha = x^\alpha \quad (\alpha = 2, \cdots, n),$$

we have for the jacobian of the transformation

$$\Delta = \left| \frac{\partial x'^i}{\partial x^j} \right| = e^\varphi .$$

* This definition is equivalent to that adopted by *Cartan*, 1924, 3. p. 223, as pointed out by *J. M. Thomas*, 1926, 3, p. 664.

Consequently

$$\frac{\partial \log \Delta}{\partial x^j} = \frac{\partial \varphi}{\partial x^j} = \Gamma^h_{hj}$$

and from (35.3) we obtain $\Gamma'^i_{ij} = 0$. Hence we have:

A necessary and sufficient condition that there exist for a given affine connection a coördinate system x^i with respect to which the connection is normal, that is

$$(37.3) \qquad \Gamma^i_{ij} = 0,$$

*is that the tensor B_{ij} be symmetric.**

Furthermore we have from (35.3):

The normal affine connection for a given coördinate system is the normal connection for all coördinate systems obtained from the given one by transformations of constant jacobian and only for these.

When equations (37.3) hold, we have from (35.2) for the normal connection

$$(37.4) \qquad \Pi^i_{jk} = \Gamma^i_{jk}.$$

Then from (35.8) we have $\Pi^h_{ijk} = B^h_{ijk}$ and from (35.14) $\Pi_{ij} = B_{ij}$. Hence the equations (35.12) become

$$(37.5) \qquad W^h_{ijk} = B^h_{ijk} + \frac{1}{n-1}(\delta^h_j B_{ik} - \delta^h_k B_{ij}).$$

Since this is the form which (32.10) assumes for a space when B_{ij} is symmetric, we have thus another proof that the tensor defined by (35.12) is the Weyl tensor.

The first theorem of this section is a corollary of the theorem:

Each of the affine connections with a given projective connection is uniquely determined by the values of Γ^i_{ij}.

In fact, from (35.2) it is seen that the coefficients of a projective connection must satisfy the conditions

$$(37.6) \qquad \Pi^i_{ij} = 0.$$

* Cf. *Cartan*, 1924, 3, p. 225 and *J. M. Thomas*, 1926, 3, p. 665.

From (35.2) it is seen also that when Γ^i_{ij} are given, Γ^i_{jk} are uniquely determined, and furthermore that the corresponding functions Γ^i_{ij} are equal to the given values of these functions.

38. Projective parameters of a path. When the expressions for Γ^i_{kl} are obtained from equations of the form (35.2) and substituted in the equations of the paths (7.6), the latter become

$$\frac{dx^j}{dt}\left(\frac{d^2x^i}{dt^2}+\Pi^i_{kl}\frac{dx^k}{dt}\frac{dx^l}{dt}\right) - \frac{dx^i}{dt}\left(\frac{d^2x^j}{dt^2}+\Pi^j_{kl}\frac{dx^k}{dt}\frac{dx^l}{dt}\right) = 0. \tag{38.1}$$

Consequently along any path we have

$$\frac{d^2x^i}{dt^2}+\Pi^i_{kl}\frac{dx^k}{dt}\frac{dx^l}{dt} = \psi\frac{dx^i}{dt}, \tag{38.2}$$

where ψ is a determinate function of t. If we define a parameter p, to within an additive constant, by the equation

$$\frac{dp}{dt} = a e^{\int \psi dt}, \tag{38.3}$$

where a is an arbitrary constant, equations (38.2) become

$$\frac{d^2x^i}{dp^2}+\Pi^i_{jk}\frac{dx^j}{dp}\frac{dx^k}{dp} = 0. \tag{38.4}$$

From the form of these equations it follows that the parameter p is not altered by a change of projective connection. It has been called a *projective* parameter of the path by T. Y. Thomas who established its existence.*

From equations (38.2) and (22.2) we have for a path in consequence of (35.2)

$$\psi = \varphi - \frac{2}{n+1}\Gamma^h_{hj}\frac{dx^j}{dt}.$$

Consequently by means of (38.3) and (22.3) we find the following relation between projective and affine parameters of a path:

* 1925, 2, p. 200; cf. also *Veblen* and *J. M. Thomas*, 1925, 4, p. 205.

$$p = b \int e^{-\frac{2}{n+1} \int \Gamma^h_{hj} \, dx^j} \, ds, \tag{38.5}$$

where b is an arbitrary constant; in the integral on the right it is understood that the x's are expressed in terms of s.

If we make use of equations (35.3) and denote by p' the parameter defined by (38.5) for a coördinate system x'^{α} we obtain the relation

$$\frac{dp}{dp'} = \frac{\bar{b}}{\Delta^{\frac{2}{n+1}}}, \tag{38.6}$$

where $\bar{b}$ is a constant. Consequently when we speak of a projective parameter it is associated with a particular coördinate system;* in this respect it is different from the affine parameter s (§ 22).

From the form of equations (38.5) and the results of § 37 we have:

A projective parameter p for a coördinate system x^i is an affine parameter of the paths for the normal connection for the x's.

When ψ_k in (32.1) are the components of the gradient of a function ψ, equation (32.3) becomes

$$\bar{s} = c \int e^{2\psi} \, ds.$$

In consequence of the results at the close of § 37 the above results are consistent with equation (38.6).

39. Coefficients of a projective connection as tensors. It is seen from equations (5.6) that the coefficients of an affine connection are components of a tensor for affine transformations of coördinates, that is,

$$x'^i = a^i_j \, x^j + b^i,$$

where the a's and b's are constant, and only for such transformations.

* Cf. *T. Y. Thomas*, 1925, 2, p. 201.

We seek the types of transformation for which the coefficients of a projective connection are components of a tensor. From (35.5) it follows that in this case we must have

$$\frac{\partial^2 x'^\alpha}{\partial x^i \partial x^j} = \frac{\partial x'^\alpha}{\partial x^j} \frac{\partial \theta}{\partial x^i} + \frac{\partial x'^\alpha}{\partial x^i} \frac{\partial \theta}{\partial x^j} \tag{39.1}$$

Expressing the conditions of integrability of these equations, we have

$$\frac{\partial x'^\alpha}{\partial x^k}\left(\frac{\partial^2 \theta}{\partial x^i \partial x^j} - \frac{\partial \theta}{\partial x^i}\frac{\partial \theta}{\partial x^j}\right) - \frac{\partial x'^\alpha}{\partial x^j}\left(\frac{\partial^2 \theta}{\partial x^i \partial x^k} - \frac{\partial \theta}{\partial x^i}\frac{\partial \theta}{\partial x^k}\right) = 0,$$

which, since they must hold for $\alpha = 1, \cdots, n$, are equivalent to

$$\frac{\partial^2 \theta}{\partial x^i \partial x^j} - \frac{\partial \theta}{\partial x^i}\frac{\partial \theta}{\partial x^j} = 0.$$

The integral of these equations is

$$e^{-\theta} = a_k x^k + h, \tag{39.2}$$

where h and the a's are arbitrary constants. When this expression is substituted in (39.1), the resulting equations can be written in the form

$$\frac{\partial}{\partial x^j}\left[(a_k x^k + h)\frac{\partial x'^\alpha}{\partial x^i}\right] + a_i \frac{\partial x'^\alpha}{\partial x^j} = 0,$$

of which the first integral is

$$(a_k x^k + h)\frac{\partial x'^\alpha}{\partial x^i} + a_i x'^\alpha = c_i^\alpha,$$

where the c's are arbitrary constants. Integrating again, we have

$$x'^\alpha = \frac{c_j^\alpha x^j + d^\alpha}{a_k x^k + h}, \tag{39.3}$$

the d's being arbitrary constants. The jacobian of this transformation is

$$(39.4) \qquad \left|\frac{\partial x'^{\alpha}}{\partial x^i}\right| = \frac{1}{(a_k x^k + h)^{n+1}} \begin{vmatrix} b & d^1 & \cdots & d^n \\ a_1 & c_1^1 & \cdots & c_1^n \\ \cdot & \cdot & \cdot & \cdot \\ \cdot & \cdot & \cdot & \cdot \\ a_n & c_n^1 & \cdots & c_n^n \end{vmatrix}^{*}$$

This result is in keeping with (39.2) and (35.6), when it is observed that in (35.6) Δ may be replaced by $c\Delta$, where c is any constant without altering (35.5). Hence we have:

The coefficients of a projective connection are components of a tensor under linear fractional transformations of the coördinates and under these alone.†

If there exists a coördinate system for a given space for which the coefficients of the projective connection Π^i_{jk} are zero, the space is projectively flat, as follows from (35.12). Conversely, when the coördinate system of a projectively flat space is such that the Γ's are given by (34.4), then $\Pi^i_{jk} = 0$. Hence we have:

A necessary and sufficient condition that a space admit a coördinate system for which its coefficients of projective connection are zero is that it be projectively flat.

When the coördinate system is such that $\Pi^i_{jk} = 0$, the most general transformation of coördinates such that $\Pi'^{\alpha}_{\beta\gamma} = 0$ is any which satisfies equations (39.1). Hence we have:

When the coördinate system of a projectively flat space is such that the coefficients of the projective connection are zero, the most general transformation of coördinates preserving this property is linear fractional.

From (38.4) it follows that in these coördinate systems the equations of the paths are

$$x^i = a^i p + b^i, \qquad x'^i = a'^i p' + b'^i,$$

where the a's and b's are constants.

When $\Pi^i_{jk} = 0$, it follows from (35.2) that

$$(39.5) \qquad \Gamma^i_{jk} = 0 \qquad (i \neq j,\ i \neq k)$$

* Cf. *Kowalewski*, 1909, 2, p. 84; *Fine*, 1905, 1, p. 505.

† Cf. *Veblen* and *J. M. Thomas*, 1926, 6, p. 284.

and that

$$\Gamma^i_{ij} = \frac{1}{n+1}\, \Gamma^h_{hj}\,(1+\delta^i_j),$$

where i is not summed. Hence we have in general

$$\Gamma^i_{ij} = \varphi_j\,(1+\delta^i_j), \tag{39.6}$$

where i and j are not summed and $\varphi_i = \dfrac{1}{n+1}\, \Gamma^h_{hi}$. Conversely, the expressions (39.5) and (39.6), in which φ_i are arbitrary functions, define the most general coefficients of affine connection for which $\Pi^i_{jk} = 0$.

40. Projective coördinates. Let P be the point of coördinates x^i_0 and consider the transformation of coördinates defined by

$$x^i = x^i_0 + \delta^i_\alpha x'^\alpha - \frac{1}{2}(\Pi^i_{\alpha\beta})_P\, x'^\alpha x'^\beta + \varphi^i, \tag{40.1}$$

where φ^i are any functions of the x''s such that they and their first and second derivatives are zero when the x''s are zero, and $(\Pi^i_{\alpha\beta})_P$ indicates the value of $\Pi^i_{\alpha\beta}$ at P. From (40.1) we have

$$\left(\frac{\partial x^i}{\partial x'^\alpha}\right)_P = \delta^i_\alpha, \qquad \left(\frac{\partial^2 x^i}{\partial x'^\alpha\, \partial x'^\beta}\right)_P = -(\Pi^i_{\alpha\beta})_P. \tag{40.2}$$

If Δ denotes the jacobian $\left|\dfrac{\partial x^i}{\partial x'^\alpha}\right|$, it follows that

$$\left(\frac{\partial \log \Delta}{\partial x'^\beta}\right)_P = \left(\frac{\partial^2 x^i}{\partial x'^\alpha\, \partial x'^\beta}\, \frac{\partial x'^\alpha}{\partial x^i}\right)_P = -(\Pi^i_{i\beta})_P = 0, \tag{40.3}$$

in consequence of (37.6). Therefore if we substitute these values in equations obtained from (35.5) by interchanging the x's and x''s, we obtain

$$(\Pi'^i_{\alpha\beta})_P = 0. \tag{40.4}$$

Hence we have:

When a transformation of coördinates of the form (40.1) *is effected, the coefficients of projective connection in the x''s are zero at the point P of coördinates x^i_0.*

We call the coördinates x'^{α} *projective coördinates.*

A particular system of projective coördinates is obtained, when we proceed with equations (38.4) in a manner analogous to that which yielded affine normal coördinates in § 23. In this case the equations of the paths through the point P of coördinates x_0^i are

$$y^i = \left(\frac{dx^i}{dp}\right)_P p, \tag{40.5}$$

where p is the projective parameter for the x's, and the equations of transformation of coördinates are

$$\begin{aligned} x^i = x_0^i + y^i - \frac{1}{2}(\Pi^i_{jk})_P\, y^j y^k \\ - \frac{1}{3!}(\Pi^i_{j_1 j_2 j_3})_P\, y^{j_1} y^{j_2} y^{j_3} + \cdots, \end{aligned} \tag{40.6}$$

where $\Pi^i_{j_1 \cdots j_r}$ are the same expressions in the Π's as (22.8) are in the Γ's.

If we denote by $\overline{\Pi}^i_{jk}$ the Π's in the y's, we have from (40.5) and the equations of the form (38.1) in the y's that the equations

$$(\overline{\Pi}^i_{kl}\, y^j - \overline{\Pi}^j_{kl}\, y^i)\, y^k y^l = 0 \tag{40.7}$$

must hold throughout the domain for which equations (40.6) define a transformation of coördinates.

From the theorem of § 38 it follows that the y's as defined by (40.5) are the affine normal coördinates corresponding to the x's for the space with normal connection for the x's. Moreover, equations (40.7) follow from (23.6) and equations of the form (35.2) for $\overline{\Pi}^i_{kl}$.

41. Projective normal coördinates. We have remarked that p in (40.5) is the projective parameter for the x's and not for the y's. We seek a system of coördinates z^i such that the equations of the paths through P (x_0^i) shall be

$$z^i = \left(\frac{dx^i}{d\overline{p}}\right)_P \overline{p}, \tag{41.1}$$

where $\bar{p}$ is the projective parameter for the z's. Moreover, we require that the z's be projective coördinates, that is,

$$(41.2) \qquad x^i = x^i_0 + z^i - \frac{1}{2}(\Pi^i_{jk})_P\, z^j z^k + \cdots,$$

where the terms of order higher than the second are as yet undetermined. From (40.6) and (41.2) we have

$$(41.3) \qquad \left(\frac{\partial z^i}{\partial x^j}\right)_P = \left(\frac{\partial y^i}{\partial x^j}\right)_P = \left(\frac{\partial y^i}{\partial z^j}\right)_P = \delta^i_j.$$

When the expressions (41.1) are substituted in equations of the form (38.4), we have

$$(41.4) \qquad P^i_{jk}\, z^{'j} z^{'k} = 0,$$

the P's being coefficients of projective connection in the z's. In order that there may exist a transformation of the y's, defined by (40.6), into z's such that (41.4) hold, we must have

$$\left(\overline{\Pi}^i_{jk}\frac{\partial z^h}{\partial y^i} - \frac{\partial^2 z^h}{\partial y^j \partial y^k} + \frac{\partial z^h}{\partial y^j}\frac{\partial \theta}{\partial y^k} + \frac{\partial z^h}{\partial y^k}\frac{\partial \theta}{\partial y^j}\right) \times \frac{\partial y^j}{\partial z^p}\frac{\partial y^k}{\partial z^q} z^{'p} z^{'q} = 0,$$

as follows from (35.5), where

$$(41.5) \qquad \theta = -\frac{1}{n+1}\log \bar{\Delta}, \qquad \bar{\Delta} = \left|\frac{\partial y^i}{\partial z^j}\right|.$$

The above equations may be written in the form

$$(41.6) \quad \left(\overline{\Pi}^i_{jk}\frac{\partial y^j}{\partial z^p}\frac{\partial y^k}{\partial z^q} + \frac{\partial^2 y^i}{\partial z^p \partial z^q} + 2\frac{\partial \theta}{\partial z^p}\frac{\partial y^i}{\partial z^q}\right) z^{'p} z^{'q} = 0.$$

Since equations (40.5) and (41.1) define the same paths, it follows that the transformation is of the form (cf. § 33)

$$(41.7) \qquad y^i = \frac{z^i}{\varphi},$$

where φ is a function whose expansion must be of the form

$$(41.8) \qquad \varphi = 1 + a_{ij}\, z^i z^j + \cdots,$$

in order that (40.6) and (41.2) be consistent. From (41.7) we have

$$\text{(41.9)} \qquad \frac{\partial y^i}{\partial z^j} = \frac{1}{\varphi^2}\left(\varphi\,\delta^i_j - z^i \frac{\partial \varphi}{\partial z^j}\right),$$

which are consistent with (41.3), in consequence of (41.8). Accordingly we have

$$\frac{\partial y^i}{\partial z^j} z^j = \frac{z^i}{\varphi^2}\left(\varphi - \frac{\partial \varphi}{\partial z^j} z^j\right)$$

$$\frac{\partial^2 y^i}{\partial z^j \partial z^k} z^j z^k = -\frac{z^i}{\varphi^3}\left[\varphi \frac{\partial^2 \varphi}{\partial z^j \partial z^k} z^j z^k + 2\left(\varphi - \frac{\partial \varphi}{\partial z^j} z^j\right)\frac{\partial \varphi}{\partial z^k} z^k\right]$$

and

$$\text{(41.10)} \qquad \bar{\Delta} = \frac{1}{\varphi^{2n}}\left|\varphi\,\delta^i_j - z^i \frac{\partial \varphi}{\partial z^j}\right| = \frac{1}{\varphi^{n+1}}\begin{vmatrix} \varphi & \frac{\partial \varphi}{\partial z^1} & \cdots & \frac{\partial \varphi}{\partial z^n} \\ z^1 & \delta^1_1 & \cdots & \delta^1_n \\ \cdot & \cdot & \cdot & \cdot \\ z^n & \delta^n_1 & \cdots & \delta^n_n \end{vmatrix}^* = \frac{1}{\varphi^{n+1}}\left(\varphi - \frac{\partial \varphi}{\partial z^j} z^j\right)$$

Consequently

$$\theta = \log \varphi - \frac{1}{n+1} \log\left(\varphi - \frac{\partial \varphi}{\partial z^j} z^j\right),$$

$$\frac{\partial \theta}{\partial z^k} z^k = \frac{1}{\varphi}\frac{\partial \varphi}{\partial z^k} z^k + \frac{\frac{1}{n+1}\frac{\partial^2 \varphi}{\partial z^j \partial z^k} z^j z^k}{\varphi - \frac{\partial \varphi}{\partial z^i} z^i}.$$

When these expressions are substituted in (41.6), we obtain

$$\text{(41.11)} \qquad \bar{\Pi}^i_{jk} \frac{z^j z^k}{\varphi^2}\left(\varphi - \frac{\partial \varphi}{\partial z^j} z^j\right)^2 + \frac{1-n}{1+n} z^i \frac{\partial^2 \varphi}{\partial z^j \partial z^k} z^j z^k = 0.$$

We remark that from (41.7) it follows that

$$\frac{\bar{\Pi}^i_{jk} z^j z^k}{z^i \varphi} = \frac{\bar{\Pi}^i_{jk} y^j y^k}{y^i},$$

* Cf. *Kowalewski*, 1909, 2, p. 84; *Fine*, 1905, 1, p. 505.

and from (40.7) that the value of the right-hand member is the same for each i, it being understood that i is not summed. From the form of equation (41.11) it is evident that it admits a solution of the form (41.8), and thus there exists a projective coördinate system z^i associated with a given coördinate system x^i, for which the equations of the paths through $P(x_0^i)$ are of the form (41.1), $\bar{p}$ being the projective parameter for the z's. Following Veblen and J. M. Thomas,* who established their existence in a different manner, we call them *projective normal coördinates.*

If instead of starting with a coördinate system x^i we had used another general coördinate system x'^i, we should have obtained another projective normal coördinate system z'^i. Then in place of (41.1) and by means of (41.1) we have

$$
\begin{aligned}
z'^i = \xi'^i \bar{p}' &= \left(\frac{dz'^i}{d\bar{p}'}\right)_0 \bar{p}' = \left(\frac{\partial z'^i}{\partial z^j}\frac{dz^j}{d\bar{p}}\frac{d\bar{p}}{d\bar{p}'}\right)_0 \bar{p}' \\
&= a^i_j \frac{z^j}{\bar{p}}\bar{p}'\left(\frac{d\bar{p}}{d\bar{p}'}\right)_0,
\end{aligned}
\tag{41.12}
$$

where $\bar{p}$ and $\bar{p}'$ are projective parameters for the respective coördinates z^i and z'^i, and in consequence of (38.6) we have

$$
\left(\frac{d\bar{p}}{d\bar{p}'}\right)_0 = \frac{1}{|a^i_j|^{\frac{2}{n+1}}}.
$$

Between the z's and z''s we have a relation of the form

$$
z'^i = \frac{a^i_j z^j}{\varphi(z)},
\tag{41.13}
$$

as follows from (41.12), where

$$
\frac{\bar{p}}{\bar{p}'} = \frac{\varphi(z)}{|a^i_j|^{\frac{2}{n+1}}}.
$$

Differentiating with respect to $\bar{p}$ and making use of (41.1), we have

* 1925, 4, p. 205.

$$\varphi - \frac{\varphi^2}{|a_j^i|^{\frac{2}{n+1}}} \frac{d\bar{p}'}{d\bar{p}} = \frac{\partial \varphi}{\partial z^k} z^k .$$

By the method used in (41.10) we find for the jacobian of the transformation (41.13)

$$\left| \frac{\partial z'^i}{\partial z^j} \right| = \frac{1}{\varphi^{n+1}} \left(\varphi - \frac{\partial \varphi}{\partial z^k} z^k \right) | a_j^i | .$$

Substituting in the preceding equation, we obtain, in consequence of (38.6),

$$\left(\varphi - \frac{\partial \varphi}{\partial z^i} z^i \right)^{\frac{1-n}{1+n}} = 1 .$$

We must find the solution of this equation such that $\left(\frac{\partial z'^i}{\partial z^j} \right)_P = a_j^i$; consequently we must have $(\varphi)_P = 1$, $\left(\frac{\partial \varphi}{\partial z^i} \right)_P = a_i$, where a_i are constants. The unique solution of this equation satisfying these conditions is $\varphi(z) = 1 + a_i z^i$. Hence we have:

When the coördinates x^i of a space undergo a general transformation, the projective normal coördinates at a point P associated with the x's undergo a linear fractional transformation

(41.14) $$z'^i = \frac{a_j^i z^j}{1 + a_k z^k} .^*$$

From (41.3) and analogous expressions in the primes and from (41.14) we have

$$\frac{\partial z^i}{\partial x^j} = \frac{\partial x'^i}{\partial z'^j} = \delta_j^i , \quad \left(\frac{\partial z'^i}{\partial z^j} \right)_P = a_j^i ,$$

and consequently

(41.15) $$\left(\frac{\partial x'^i}{\partial x^j} \right)_P = a_j^i .$$

* This result has been established in a different manner by *Veblen* and *J. M. Thomas*, 1925, 4, p. 206.

Also from the law of multiplication of jacobians, from equations of the form (40.3) for transformations of the form (40.1) and from (41.13) we have

$$(41.16)\quad \left(\frac{\partial}{\partial x^k}\log\left|\frac{\partial x'^i}{\partial x^j}\right|\right)_P = \left(\frac{\partial}{\partial z^k}\log\left|\frac{\partial z'^i}{\partial z^j}\right|\right)_P = -(n+1)\,a_k.$$

Consequently equations (41.15) and (41.16) give the significance of the constants in (41.14) and the original transformation in the x's and x''s.

42. Significance of a projective change of affine connection. In consequence of equations (40.2), (40.3) and (35.3) we have for any system of projective coördinates associated with a coördinate system x^i and with $P(x^i_0)$ for origin

$$(42.1)\qquad (\Gamma^i_{ij})_P = (C^i_{ij})_P,$$

where C^i_{jk} are the coefficients of affine connection in these projective coördinates. Suppose that the latter are the projective normal coördinates z^i associated with the x's. If we introduce homogeneous coördinates, putting

$$(42.2)\qquad z^i = \frac{\bar{z}^i}{\bar{z}^0},$$

equations (41.14) become

$$(42.3)\qquad \bar{z}'^\alpha = b^\alpha_\beta\,\bar{z}^\beta,$$

where

$$(42.4)\qquad b^i_j = a^i_j,\quad b^i_0 = 0,\quad b^0_0 = 1,\quad b^0_i = a_i.^*$$

If we put

$$(42.5)\quad (C^i_{ij})_P = -(n+1)u_j,\ (C'^i_{ij})_P = -(n+1)u'_j,\ u_0 = u'_0 = 1,$$

from equations of the form (35.3) in the z's and z''s we have, in consequence of (41.16) and (42.3),

$$\bar{u}'_\alpha\,\bar{z}'^\alpha = \bar{u}_\beta\,z^\beta.$$

Hence the u's are transformed contragrediently to the z's.

* Here it is understood that greek indices take the values $0, 1, \cdots, n$ and latin $1, \cdots, n$.

Accordingly if at P we look upon the $\bar{z}$'s as homogeneous coördinates of a projective space, each choice of an affine connection singles out a hyperplane, which justifies the use of the term affine.

In § 37 we saw that among all the spaces with the same projective connection there is one for which $\Gamma^i_{ij} = 0$ in the given coördinate system x^i. From (42.1) and (42.5) it is seen that at every point in the coördinate system z^i at the point associated with the x's we have $u_j = 0$ $(j = 1, \cdots, n)$. Consequently in this coördinate system the plane at infinity is $\bar{z}^0 = 0$. Accordingly at each point the $\bar{z}$'s are homogeneous cartesian coördinates, defined by (42.2) in which the z's are cartesian.*

43. Homogeneous first integrals under a projective change. If for a given affine connection Γ^i_{jk} the equations of the paths (22.4) admit the homogeneous first integral

$$a_{r_1 \cdots r_m} \frac{dx^{r_1}}{ds} \cdots \frac{dx^{r_m}}{ds} = \text{const.},$$

it follows from (32.3) that for a projective change of connection defined by (32.1), the equations (32.2) of the paths admit the first integral

$$e^{2m\int \psi_k dx^k} a_{r_1 \cdots r_m} \frac{dx^{r_1}}{d\bar{s}} \cdots \frac{dx^{r_m}}{d\bar{s}} = \text{const.},$$

where the integral $\int \psi_k \, dx^k$ is taken along the path in question.

Conversely, if the equations (22.4) of the paths admit a first integral

$$e^{\int \varphi_k dx^k} a_{r_1 \cdots r_m} \frac{dx^{r_1}}{ds} \cdots \frac{dx^{r_m}}{ds} = \text{const.}, \tag{43.1}$$

then for the affine connection defined by (32.1), in which

$$\psi_i = -\frac{\varphi_i}{2m}, \tag{43.2}$$

* Cf. *Veblen* and *J. M. Thomas*, 1926, 6, p. 295.

the corresponding equations (32.2) admit the first integral

$$a_{r_1 \cdots r_m} \frac{d x^{r_1}}{d \bar{s}} \cdots \frac{d x^{r_m}}{d \bar{s}} = \text{const.} \tag{43.3}$$

When we express the condition that (43.1) be a first integral of equations (22.4), we get

$$P(a_{r_1 \cdots r_m, k} + a_{r_1 \cdots r_m} \varphi_k) = 0, \tag{43.4}$$

where P denotes the sum of the terms obtained from those in parenthesis by cyclic permutation of the indices (cf. § 31). Hence we have:

A necessary and sufficient condition that a projective change of affine connection can be effected so that the corresponding equations of the paths shall admit a homogeneous first integral of the mth degree is the existence of a symmetric tensor $a_{r_1 \cdots r_m}$ *and a vector* φ_i *such that equations* (43.4) *are satisfied; then the projective change is defined by* (32.1) *in which* $\psi_i = -\varphi_i/2m$ *and the first integral is given by* (43.3).*

When, and only when, φ_k in (43.1) is a gradient, equation (43.1) is of the form (43.3) Hence the results of § 31 may be stated as follows:

A necessary and sufficient condition that the equations of the paths for a given affine connection admit a homogeneous integral of the mth degree is that there exist a tensor $a_{r_1 \cdots r_m}$ *and a gradient* $\varphi_{,k}$ *such that the corresponding equations* (43.4) *hold; then the first integral is*

$$e^{\varphi} a_{r_1 \cdots r_m} \frac{d x^{r_1}}{d s} \cdots \frac{d x^{r_m}}{d s} = \text{const.}\dagger$$

The condition (43.4) is satisfied by the tensor g_{ij} and the vector φ_i of a Weyl geometry, as follows from (30.1). Consequently the equations of the paths for the given affine connection admit the first integral

$$e^{2\int \varphi_k dx^k} g_{ij} \frac{dx^i}{ds} \frac{dx^j}{ds} = \text{const.}$$

* Cf. *Eisenhart,* 1924, 2, p. 381; also *J. M. Thomas,* 1926, 7, p. 119.

† Cf. *Veblen* and *T. Y. Thomas,* 1923, 1, p. 583.

From (43.2) it follows that for the projective change defined by (32.1) in which $\psi_i = -\varphi_i/2$ the new affine connection is such that its equations of the paths admit the first integral

$$(43.5) \qquad g_{ij} \frac{dx^i}{ds} \frac{dx^j}{ds} = \text{const.}$$

If $g_{ij,\bar{k}}$ denotes the covariant derivative with respect to the new connection, from (30.1) we have

$$(43.6) \qquad g_{ij,\bar{k}} + g_{jk}\, \psi_i + g_{ik}\, \psi_j - 2 g_{ij}\, \psi_k = 0.$$

From these equations we have

$$(43.7) \qquad g^{ij}\, g_{ij,\bar{k}} = 2(n-1)\, \psi_k,$$

and consequently (43.6) can be written as

$$(43.8) \quad 2(n-1)\, g_{ij,\bar{k}} + g^{pq}(g_{jk}\, g_{pq,\bar{i}} + g_{ik}\, g_{pq,\bar{j}} - 2 g_{ij}\, g_{pq,\bar{k}}) = 0.$$

Conversely, if the equations of the paths of an affine connection $\bar{\Gamma}^i_{jk}$ admit a first integral (43.5) and equations (43.8) are satisfied, for the vector ψ_k defined by (43.7) equations (43.8) reduce to (43.6) and by means of (32.1) with $\varphi_i = -2\psi_i$, we get (30.1). Hence we have the following theorem due to J. M. Thomas:*

A necessary and sufficient condition that an affine geometry whose paths admit a quadratic integral (43.5) *have the same paths as a Weyl geometry is that equations* (43.8) *be satisfied, covariant differentiation being with respect to the given connection.*

44. Spaces for which the equations of the paths admit $n(n+1)/2$ independent homogeneous linear first integrals. In order that the equations (22.4) of the paths admit a linear first integral

$$(44.1) \qquad a_i \frac{dx^i}{ds} = \text{const.},$$

it is necessary that (§ 31)

$$(44.2) \qquad a_{i,j} + a_{j,i} = 0.$$

* 1926, 7, p. 122.

Differentiating covariantly with respect to x^k, we have

$$a_{i,jk} + a_{j,ik} = 0. \tag{44.3}$$

If from the sum of this equation and the first of the following

$$a_{k,ij} + a_{i,kj} = 0, \qquad a_{j,ki} + a_{k,ji} = 0$$

we subtract the second, the resulting equation is reducible by means of (6.4), (21.3) and (21.4) to

$$a_{i,jk} = -a_l B^l_{kij}. \tag{44.4}$$

The conditions of integrability of these equations are reducible to

$$\begin{aligned} &a_h (B^h_{kij,l} - B^h_{lij,k}) \\ &+ a_{h,p} (\delta^p_l B^h_{kij} - \delta^p_k B^h_{lij} + \delta^p_j B^h_{ikl} - \delta^p_i B^h_{jkl}) = 0. \end{aligned} \tag{44.5}$$

If we put

$$\frac{\partial a_i}{\partial x^j} = a_h \Gamma^h_{ij} + b_{ij},$$

these equations and (44.4), written as

$$b_{ij,k} = -a_l B^l_{kij},$$

constitute a system of equations of the form (8.1) in the n quantities a_i and the n^2 quantities b_{ij}. In this case we have a system F_0 of $n(n+1)/2$ equations $b_{ij} + b_{ji} = 0$ which follow from (44.2). Equations (44.5) are the set F_1 for this case. Hence we can apply the results of § 8 to get the conditions to be satisfied, in order that there be one or more first integrals.* When (44.5) are satisfied identically in virtue of (44.2), the solution admits $n(n+1)/2$ arbitrary constants; it is this case we consider in what follows.

From (44.5) we have the following equations of condition:

$$B^h_{kij,l} - B^h_{lij,k} = 0, \tag{44.6}$$

* Cf. *Veblen* and *T. Y. Thomas*, 1923, 1, pp. 591–599.

$$\begin{aligned}(44.7)\qquad \delta_l^p B^h_{kij} - \delta_l^h B^p_{kij} - \delta_k^p B^h_{lij} + \delta_k^h B^p_{lij} + \delta_j^p B^h_{ikl} - \delta_j^h B^p_{ikl} \\ - \delta_i^p B^h_{jkl} + \delta_i^h B^p_{jkl} = 0.\end{aligned}$$

Contracting (44.7) for p and l, we obtain (cf. § 5)

$$(44.8)\qquad B^h_{kij} = \frac{1}{n-1}(\delta^h_j B_{ik} - \delta^h_i B_{jk} + 2\delta^h_k \beta_{ij}).$$

Contracting this equation for h and j, we have

$$B_{ki} - B_{ik} = \frac{2}{n-1}\beta_{ik}.$$

Since β_{ik} is the skew-symmetric part of B_{ik} (§ 5), it follows from these equations that $\beta_{ij} = 0$, and consequently B_{ij} is symmetric. Hence (44.8) reduce to

$$(44.9)\qquad B^h_{kij} = \frac{1}{n-1}(\delta^h_j B_{ik} - \delta^h_i B_{jk}).$$

When these expressions are substituted in (44.7), we find that these conditions are satisfied identically. Again when they are substituted in (44.6), we obtain

$$(44.10)\qquad B_{ik,l} - B_{il,k} = 0.$$

Comparing these results with those of § 34, we have:

A necessary and sufficient condition that the equations of the paths of a space admit homogeneous linear first integrals involving $n(n+1)/2$ arbitrary constants is that the space be projectively flat and the tensor B_{ij} be symmetric.

In § 34 it was seen that any such space is determined by taking

$$(44.11)\qquad \Gamma^i_{jk} = -(\delta^i_j \psi_{,k} + \delta^i_k \psi_{,j}),$$

where $\psi_{,j}$ is an arbitrary gradient, and that the coördinate system x^i for which the Γ's have this form is cartesian in the corresponding flat space. In this coördinate system and for the Γ's given by (44.11) equations (44.2) become

$$(44.12)\qquad \frac{\partial b_i}{\partial x^j} + \frac{\partial b_j}{\partial x^i} = 0.$$

where

$$b_i = a_i e^{2\psi}. \tag{44.13}$$

Equations (44.12) are the form which (44.2) assume in a flat space referred to cartesian coördinates. In this case equations (44.4) become

$$\frac{\partial^2 b_i}{\partial x^j \partial x^k} = 0. \tag{44.14}$$

From (44.12) for $i = j$ it follows that b_i is independent of x^i. Then from (44.12) and (44.14) it follows that the general solution is

$$b_i = c_{ij} x^j + d_i, \tag{44.15}$$

where c_{ij} and d_i are constants, subject only to the condition that c_{ij} is skew-symmetric in the indices. Hence there are $n(n+1)/2$ arbitrary constants, as desired, and for the given space a_i are given by (44.13) and (44.15).*

45. Transformations of the equations of the paths. Equations (35.5) may be obtained in another manner. In § 22 it was shown that the affine parameter of a path is not changed by a general transformation of coördinates. Conversely, if we take the equations of the paths in the form (22.4) in two coördinate systems x^i and x'^i and assume that s is unaltered by the transformation, we obtain (5.6). We wish now to consider the more general case when s is not invariant under the change of coördinates. To this end we take the equations of the paths in the form (7.6), and seek the conditions which the Γ's and Γ''s must satisfy in order that (7.6) are transformable into

$$\begin{aligned} &\frac{dx'^\alpha}{dt}\left(\frac{d^2x'^\beta}{dt^2} + \Gamma'^\beta_{\gamma\delta}\frac{dx'^\gamma}{dt}\frac{dx'^\delta}{dt}\right) \\ &\qquad - \frac{dx'^\beta}{dt}\left(\frac{d^2x'^\alpha}{dt^2} + \Gamma'^\alpha_{\gamma\delta}\frac{dx'^\gamma}{dt}\frac{dx'^\delta}{dt}\right) = 0 \end{aligned} \tag{45.1}$$

by a change of coördinates.

* Cf. *Eisenhart*, 1926, 9, p. 336.

If we effect the transformation $x'^{\alpha} = \varphi^{\alpha}(x^1, \cdots, x^n)$ on (45.1) and express the condition that (7.6) be satisfied, we obtain

$$\left(\frac{\partial x'^{\alpha}}{\partial x^k} A^{\beta}_{ij} - \frac{\partial x'^{\beta}}{\partial x^k} A^{\alpha}_{.ij}\right) \frac{dx^i}{dt} \frac{dx^j}{dt} \frac{dx^k}{dt} = 0,$$

where

$$(45.2)\quad A^{\alpha}_{ij} = \frac{\partial^2 x'^{\alpha}}{\partial x^i \partial x^j} + \Gamma'^{\alpha}_{\beta\gamma} \frac{\partial x'^{\beta}}{\partial x^i} \frac{\partial x'^{\gamma}}{\partial x^j} - \Gamma^k_{ij} \frac{\partial x'^{\alpha}}{\partial x^k}.$$

Since the above conditions must be satisfied for all the paths, we must have

$$(45.3)\quad \begin{aligned} &\frac{\partial x'^{\alpha}}{\partial x^k} A^{\beta}_{ij} - \frac{\partial x'^{\beta}}{\partial x^k} A^{\alpha}_{ij} + \frac{\partial x'^{\alpha}}{\partial x^i} A^{\beta}_{jk} \\ &\qquad - \frac{\partial x'^{\beta}}{\partial x^i} A^{\alpha}_{jk} + \frac{\partial x'^{\alpha}}{\partial x^j} A^{\beta}_{ki} - \frac{\partial x'^{\beta}}{\partial x^j} A^{\alpha}_{ki} = 0. \end{aligned}$$

Multiplying by $\dfrac{\partial x^k}{\partial x'^{\alpha}}$ and summing for k and α, we obtain in consequence of (45.2)

$$(45.4)\quad \begin{aligned} (n+1)\, A^{\beta}_{ij} &= \frac{\partial x'^{\beta}}{\partial x^j} \left(\frac{\partial \log \Delta}{\partial x^i} + \Gamma'^{\alpha}_{\alpha\gamma} \frac{\partial x'^{\gamma}}{\partial x^i} - \Gamma^k_{ki}\right) \\ &\quad + \frac{\partial x'^{\beta}}{\partial x^i} \left(\frac{\partial \log \Delta}{\partial x^j} + \Gamma'^{\alpha}_{\alpha\gamma} \frac{\partial x'^{\gamma}}{\partial x^j} - \Gamma^k_{kj}\right), \end{aligned}$$

where Δ is the jacobian

$$(45.5)\quad \Delta = \left|\frac{\partial x'^{\alpha}}{\partial x^i}\right|.$$

When the expressions (45.4) are substituted in (45.3), the latter are satisfied identically. Hence the conditions are given by combining (45.2) and (45.4); this gives equations (35.5).

From (22.2) and (22.3) we have

$$(45.6)\quad \frac{d^2 x^i}{dt^2} + \Gamma^i_{jk} \frac{dx^j}{dt} \frac{dx^k}{dt} = \frac{\dfrac{d^2 s}{dt^2}}{\dfrac{ds}{dt}} \frac{dx^i}{dt}$$

for the determination of the affine parameter s of a path. By means of (35.5) we obtain from (45.6) and analogous equations in the Γ''s and the affine parameter s'

$$
\frac{\dfrac{d^2 s'}{dt^2}}{\dfrac{ds'}{dt}} = \frac{\dfrac{d^2 s}{dt^2}}{\dfrac{ds}{dt}} + \frac{2}{n+1}\left(\Gamma'^{\beta}_{\beta\gamma}\frac{\partial x'^{\gamma}}{\partial x^k} - \Gamma^h_{hk}\right)\frac{dx^k}{dt} + \frac{2}{n+1}\frac{d\log\Delta}{dt} \tag{45.7}
$$

If the affine parameters s and s' are to be equal, we must have

$$
\Gamma^h_{hk} = \Gamma'^{\beta}_{\beta\gamma}\frac{\partial x'^{\gamma}}{\partial x^k} + \frac{\partial \log \Delta}{\partial x^k} \tag{45.8}
$$

in which case equations (35.5) reduce to equations analogous to (5.6); we have seen that equations (45.8) are a consequence of these.

If we consider the most general solution of equations (35.5), when the coördinates are not changed but only the affine parameter, we have $\Pi'^i_{jk} = \Pi^i_{jk}$, which shows the invariant character of the Π's under a projective change.

46. Collineations in an affinely connected space. The results of § 45 may be used to define transformations of points of an affinely connected manifold into points of the manifold such that paths are transformed into paths. We call such transformations *collineations*. The conditions to be satisfied by a space in order that it may admit one or more collineations arise from equations (35.5) on the assumption that each pair of coefficients Γ^i_{jk} and Γ'^i_{jk} with the same indices are the same functions of x^i and x'^i respectively. This is a particular case of the problem considered in § 36 and may be handled in that manner. However, if the finite equations of the transformation involve r ($\geqq 1$) parameters and possess the group property, they define a finite continuous group of collineations. In this case the transformations may be considered as generated by r infinitesimal transformations.* Accordingly we consider infinitesimal collineations as the basis of another method of obtaining affinely connected spaces which admit collineations.

An infinitesimal transformation is defined by

$$x'^i = x^i + \xi^i \delta u, \tag{46.1}$$

where ξ^i are functions of the x's and δu is an infinitesimal. Since by hypothesis the Γ's and Γ''s with the same indices are the same functions of the x's and x''s respectively, the same is true of the Π's and Π''s, as defined by (35.2); hence by Taylor's expansion we have

$$\Pi'^i_{jk} = \Pi^i_{jk} + \frac{\partial \Pi^i_{jk}}{\partial x^h} \xi^h \delta u, \tag{46.2}$$

neglecting infinitesimals of the second and higher orders; this will be done in what follows. From (46.1) it follows that the determinant Δ of the transformation is given by

$$\Delta = 1 + \frac{\partial \xi^h}{\partial x^h} \delta u,$$

and consequently

$$\frac{\partial \log \Delta}{\partial x^i} = \frac{\partial^2 \xi^h}{\partial x^h \partial x^i} \delta u. \tag{46.3}$$

When these values are substituted in (35.5), we obtain, on neglecting the multipler δu,

$$\begin{aligned} &\frac{\partial^2 \xi^\alpha}{\partial x^i \partial x^j} + \Pi^\alpha_{ik} \frac{\partial \xi^k}{\partial x^j} + \Pi^\alpha_{jk} \frac{\partial \xi^k}{\partial x^i} + \xi^h \frac{\partial \Pi^\alpha_{ij}}{\partial x^h} - \Pi^k_{ij} \frac{\partial \xi^\alpha}{\partial x^k} \\ &\quad - \frac{1}{n+1} \left(\delta^\alpha_j \frac{\partial^2 \xi^h}{\partial x^h \partial x^i} + \delta^\alpha_i \frac{\partial^2 \xi^h}{\partial x^h \partial x^j} \right) = 0. \end{aligned} \tag{46.4}$$

Because of (35.2) these equations are equivalent to

$$\begin{aligned} &\frac{\partial^2 \xi^\alpha}{\partial x^i \partial x^j} + \Gamma^\alpha_{ik} \frac{\partial \xi^k}{\partial x^j} + \Gamma^\alpha_{jk} \frac{\partial \xi^k}{\partial x^i} + \xi^h \frac{\partial \Gamma^\alpha_{ij}}{\partial x^h} - \Gamma^k_{ij} \frac{\partial \xi^\alpha}{\partial x^k} \\ &\quad = \delta^\alpha_j \varphi_i + \delta^\alpha_i \varphi_j, \end{aligned} \tag{46.5}$$

* The reader is supposed to be conversant with the Lie theory of groups as contained in the treatise of *Lie*, 1893, 1 or *Bianchi*, 1918, 1; a résumé of this theory is given by the author, 1926, 1, pp. 221–227.

where by contraction we have

$$(46.6)\qquad \varphi_i = \frac{1}{n+1}\left(\frac{\partial^2 \xi^h}{\partial x^h\,\partial x^i} + \Gamma^h_{hk}\frac{\partial \xi^k}{\partial x^i} + \xi^h \frac{\partial \Gamma^k_{ki}}{\partial x^h}\right).$$

Equations (46.5) may be written in the form

$$(46.7)\qquad \xi^h{}_{,ij} = \xi^k B^h_{ijk} + \delta^h_j\,\varphi_i + \delta^h_i\,\varphi_j.$$

Contracting for h and i, we have

$$(46.8)\qquad \xi^i{}_{,ij} = \xi^k S_{jk} + (1+n)\,\varphi_j.$$

In order that the affine parameter s be unaltered by the infinitesimal transformation (46.1), it follows from (45.8) and (46.3) that φ_i as defined by (46.6) are zero. In this case equations (46.7) become

$$(46.9)\qquad \xi^h{}_{,ij} = \xi^k B^h_{ijk}.$$

Hence when a set of functions ξ^i are a solution of (46.9) equations (46.1) define an infinitesimal collineation which preserves the affine properties of the space, and when they are a solution of (46.7), where $\varphi_i \neq 0$, the collineation preserves the projective properties. Accordingly we call them infinitesimal *affine* and *projective* collineations respectively.*

Consider the case of a projectively flat space and assume that the coördinates x^i are such that $\Pi^i_{jk} = 0$ (cf. § 39). Under these conditions equations (46.4) may be written

$$(46.10)\qquad \frac{\partial^2 \xi^\alpha}{\partial x^i\,\partial x^j} = \delta^\alpha_j\,\varphi_i + \delta^\alpha_i\,\varphi_j.$$

The conditions of integrability of these equations are

$$\delta^\alpha_j \frac{\partial \varphi_i}{\partial x^k} + \delta^\alpha_i \frac{\partial \varphi_j}{\partial x^k} = \delta^\alpha_k \frac{\partial \varphi_i}{\partial x^j} + \delta^\alpha_i \frac{\partial \varphi_k}{\partial x^j}.$$

Contracting for α and i, we find that φ_i is the gradient of a function φ, that is, $\varphi_i = \dfrac{\partial \varphi}{\partial x^i}$. Substituting in the preceding equations, we have $\dfrac{\partial^2 \varphi}{\partial x^i\,\partial x^j} = 0$ and consequently

* Cf. *Eisenhart* and *Knebelman*, 1927, 2.

$$\varphi = a_h x^h + d,$$

where the a's and d are arbitrary constants. Then from (46.10) we have

$$\xi^i = a_h x^h x^i + b^i_l x^l + c^i,$$

where the b's and c's are arbitrary constants. We recognize these expressions as defining the most general infinitesimal projective collineation in a projectively flat space.* If, on the other hand, we consider equations (46.9) for a flat space referred to cartesian coördinates, we have $\frac{\partial^2 \xi^\alpha}{\partial x^i \partial x^j} = 0$ and consequently

$$\xi^i = a^i_h x^h + b^i,$$

which define the most general infinitesimal affine collineation.† Thus as defined affine and projective infinitesimal collineations are generalizations of these respective collineations of a flat space.

Suppose that we have a solution ξ^i of equations (46.7) and that the coördinates x^i are chosen so that in this coördinate system

(46.11) $$\xi^1 = 1, \quad \xi^\alpha = 0 \quad (\alpha = 2, \cdots, n).‡$$

In this case equations (46.4) reduce to

(46.12) $$\frac{\partial \Pi^i_{jk}}{\partial x^1} = 0.$$

By means of these equations we shall prove the theorem:

When an affinely connected space admits an infinitesimal projective or affine collineation, the transformations of the finite group G_1 generated by it are collineations.

In fact, for the chosen coördinate system the equations of the finite group are

(46.13) $$x'^1 = x^1 + a, \quad x'^\alpha = x^\alpha \quad (\alpha = 2, \cdots, n),$$

* Cf. *Lie*, 1893, 1, p. 24.

† Cf. *Lie*, l. c., p. 85.

‡ 1926, 1, p. 223.

where a is a parameter. For this transformation equations (35.5) reduce to $\Pi'^{i}_{jk} = \Pi^{i}_{jk}$. In consequence of (46.12) this condition is satisfied for a projective collineation. For an affine collineation (46.11) we have from (46.5) that Γ^{i}_{jk} is independent of x^1, so that the theorem follows in this case also. Moreover, we have shown incidentally that

The most general affinely connected manifold which admits a finite group G_1 of affine collineations is given by taking for Γ^{i}_{jk} functions of $n-1$ of the coördinates.

In consequence of (39.5) and (39.6), we have that equations (46.12) are equivalent to

$$\frac{\partial \Gamma^{i}_{jk}}{\partial x^1} = 0 \qquad (i \neq j,\ i \neq k)$$

and

$$\frac{\partial \Gamma^{i}_{ii}}{\partial x^1} = 2\varphi_i, \qquad \frac{\partial \Gamma^{\alpha}_{\alpha i}}{\partial x^1} = \varphi_i$$
$$(i = 1, \cdots, n;\ \alpha = 1, \cdots, n;\ \alpha \neq i),$$

where i and α are not summed. By means of these equations we are in a position to choose the coefficients Γ^{i}_{jk} of an affine connection so that the manifold shall admit a group G_1 of projective collineations. This result is seen also from (46.5).

If ξ^i is a solution of equations (46.7) for a given connected manifold, it follows from (46.4) that it defines a collineation for every manifold in projective correspondence with the given manifold. If the coefficients of any such manifold are given by (35.1), we have

$$\text{(46.14)} \qquad \overline{\Gamma}^{h}_{hk} = \Gamma^{h}_{hk} + (n+1)\,\psi_k,$$

and consequently from (46.6) we have that the functions $\overline{\varphi}_i$ in this case are given by

$$\text{(46.15)} \qquad \overline{\varphi}_i = \varphi_i + \psi_k \frac{\partial \xi^k}{\partial x^i} + \xi^h \frac{\partial \psi_i}{\partial x^h}.$$

If we denote by $\xi^h_{,\overline{ij}}$ the second covariant derivative of ξ^h with respect to the $\overline{\Gamma}$'s, we have

$$\xi^h_{,\overline{ij}} = \xi^h_{,ij} + \delta^h_i (\psi_k \xi^k)_{,j} + \delta^h_j (\xi^k_{,i} \psi_k + \xi^k \psi_k \psi_i) \\ + \xi^h (\psi_{i,j} - \psi_i \psi_j).$$

In consequence of (32.4) these expressions and (46.15) satisfy equations of the form (46.7).

From (46.15) it is seen that the collineation determined by ξ^i is projective for the connection of coefficients $\overline{\Gamma}^i_{jk}$ unless ψ_i satisfies the conditions

$$\xi^h \frac{\partial \psi_i}{\partial x^h} + \psi_k \frac{\partial \xi^k}{\partial x^i} + \varphi_i = 0.$$

When the coördinates x^i are chosen so that ξ^i have the values (46.11), these equations become

$$\frac{\partial \psi_i}{\partial x^1} + \varphi_i = 0.$$

The general solution of these equations involves n arbitrary functions of $x^2, \cdots, x^n$. Hence we have:

When a projective or affine collineation of an affinely connected manifold is known, it is an affine collineation of a sub-group of connections projectively related to the given connection; the determination of the sub-group involves n arbitrary functions whose jacobian is zero.

47. Conditions for the existence of infinitesimal collineations. If we make use of the Ricci identities (§ 6)

$$\xi^h_{,ijk} - \xi^h_{,ikj} = \xi^h_{,r} B^r_{ijk} - \xi^l_{,i} B^h_{ljk},$$

the conditions of integrability of equations (46.7) are reducible by means of the Bianchi identities (§ 21) to

$$(47.1) \quad \begin{aligned} &\xi^l B^h_{ijk,l} + \xi^l_{,k} B^h_{ijl} - \xi^l_{,j} B^h_{ikl} + \xi^l_{,i} B^h_{ljk} - \xi^h_{,r} B^r_{ijk} \\ &\quad + \delta^h_j \varphi_{i,k} - \delta^h_k \varphi_{i,j} + \delta^h_i (\varphi_{j,k} - \varphi_{k,j}) = 0. \end{aligned}$$

Contracting for h and i, and h and k we have (cf. § 5) respectively

$$(47.2) \quad \xi^l S_{jk,l} + \xi^l_{,k} S_{jl} - \xi^l_{,j} S_{kl} + (n+1)(\varphi_{j,k} - \varphi_{k,j}) = 0,$$

$$(47.3)\quad \begin{aligned}\xi^l B_{ij,l} + \xi^l_{,j} B_{il} + \xi^l_{,i} B_{lj} + (1-n)\varphi_{i,j} \\ + (\varphi_{j,i} - \varphi_{i,j}) = 0.\end{aligned}$$

Interchange i and j in (47.3) and subtract from (47.3); in consequence of § 5 the resulting equations are of the form (47.2). From (47.2) and (47.3) we obtain

$$(47.4)\qquad \varphi_{i,j} = \frac{1}{n-1}(\xi^l C_{ij,l} + \xi^l_{,j} C_{il} + \xi^l_{,i} C_{lj}),$$

where

$$(47.5)\qquad C_{ij} = B_{ij} + \frac{1}{n+1} S_{ij}.$$

If in (47.1) we substitute for the covariant derivatives of φ_i their expressions of the form (47.4), the resulting equations are reducible by means of (32.10) and the results of § 5 to

$$(47.6)\quad \xi^l W^h_{ijk,l} + \xi^l_{,k} W^h_{ijl} - \xi^l_{,j} W^h_{ikl} + \xi^l_{,i} W^h_{ljk} - \xi^h_{,l} W^l_{ijk} = 0.$$

The conditions of integrability of equations (47.4) are obtained from the identities $\varphi_{i,jk} - \varphi_{i,kj} = \varphi_h B^h_{ijk}$. In consequence of the results of § 5, they are reducible to

$$\begin{aligned}\xi^l (C_{ij,lk} - C_{ik,lj} - C_{ih} B^h_{ljk} + C_{hj} B^h_{ikl} - C_{hk} B^h_{ijl}) + \xi^l_{,k}(C_{ij,l} - C_{il,j}) \\ + \xi^l_{,j}(C_{il,k} - C_{ik,l}) + \xi^l_{,i}(C_{lj,k} - C_{lk,j}) + (1-n)\varphi_h W^h_{ijk} = 0.\end{aligned}$$

Because of the Ricci identities for $C_{ij,kl}$ and $C_{ik,lj}$ and the identities (21.4), these equations are equivalent to

$$(47.7)\quad \begin{aligned}\xi^l (C_{ij,kl} - C_{ik,jl}) + \xi^l_{,k}(C_{ij,l} - C_{il,j}) + \xi^l_{,j}(C_{il,k} - C_{ik,l}) \\ + \xi^l_{,i}(C_{lj,k} - C_{lk,j}) + (1-n)\varphi_h W^h_{ijk} = 0.\end{aligned}$$

If equations (47.6) are differentiated covariantly and use is made of the Ricci identities for $W^h_{ijk,lr}$, the resulting equations are reducible to

$$(47.8)\quad \begin{aligned}\xi^l W^h_{ijk,rl} + \xi^l_{,r} W^h_{ijk,l} + \xi^l_{,k} W^h_{ijl,r} - \xi^l_{,j} W^h_{ikl,r} \\ + \xi^l_{,i} W^h_{ljk,r} - \xi^h_{,l} W^l_{ijk,r} + W^h_{ijr}\varphi_k - W^h_{ikr}\varphi_j \\ + W^h_{rjk}\varphi_i + 2 W^h_{ijk}\varphi_r - \delta^h_r W^l_{ijk}\varphi_l = 0.\end{aligned}$$

Contracting for h and r, we obtain

$$(47.9)\qquad \begin{aligned} &\xi^l W^h_{ijk,hl} + \xi^l_{,k} W^h_{ijl,h} - \xi^l_{,j} W^h_{ikl,h} \\ &\qquad + \xi^l_{,i} W^h_{ljk,h} + (2-n)\varphi_h W^h_{ijk} = 0. \end{aligned}$$

If we put

$$(47.10)\qquad \xi^h_{,i} = \eta^h_i,$$

equations (46.7) become

$$(47.11)\qquad \eta^h_{i,j} = \xi^k B^h_{ijk} + \delta^h_j \varphi_i + \delta^h_i \varphi_j.$$

These two sets of equations and the set (47.4) are of the form (8.1) in the functions ξ^h, η^h_i and φ_i. By means of (47.11) and (21.4) the conditions of integrability of (47.10) are satisfied identically. The conditions of integrability of (47.11) and (47.4) are given by (47.6) and (47.7) which together constitute the set F_1 of the theorem of § 8. However, when $n > 2$, equations (47.7) and (47.9) are equivalent in consequence of (32.16), which may be written

$$W^i_{jkl,i} = \frac{n-2}{n-1}(C_{jk,l} - C_{jl,k}).$$

Hence as observed in § 8 we may apply the theorem to this case taking (47.6) as the set F_1, (47.8) as the set F_2 and so on. Since all of the equations are linear and homogeneous in the dependent functions, we have:

A necessary and sufficient condition that an affinely connected space for $n > 2$ admit r ($\geqq 1$) infinitesimal projective collineations is that there exist two positive integers N and r such that the matrices of the equations $F_1, \cdots, F_N$ and $F_1, \cdots, F_{N+1}$ are of rank $n^2 + 2n - r$; when $r = 1$, the solution involves a quadrature; when $r > 1$, the general solution is a linear function with constant coefficients of r fundamental sets of solutions.

When $n = 2$, the Weyl tensor vanishes identically, and consequently equations (47.6). The above theorem applies to this case with the understanding that equations (47.7) with $W^h_{ijk} = 0$ constitute the set F_1, and the other sets are derived from this one.

From the form of equations (47.6) and (47.7) and the results of § 34 we have:

The maximum number of independent infinitesimal projective collineations which a space can admit is n^2+2n; *this is the case when, and only when, the space is projectively flat.*

The determination of spaces admitting infinitesimal affine collineations reduces to the solution of equations (47.10) and (47.11) in which $\varphi_i = 0$. In this case we have a theorem analogous to the first of the above theorems for which the sets F_1 and F_2 are obtained from (47.1) by putting $\varphi_i = 0$ and from (47.8) by replacing W^h_{ijk} by B^h_{ijk}. Since there are n^2+n functions ξ^h and η^h_i in this case, we have:

The maximum number of independent infinitesimal affine collineations which a space can admit is n^2+n; *this is the case when, and only when, the space is flat.*

The forms of the solutions ξ^i for projectively flat and flat spaces in cartesian coördinates have been obtained in § 46.

A special type of collineations is that for which the path curves of the collineations, namely the congruence determined by ξ^i, are paths of the manifold. In this case the functions ξ^i must satisfy the conditions (cf. (7.5))

$$\xi^k(\xi^j\xi^i{}_{,k}-\xi^i\xi^j{}_{,k}) = 0. \tag{47.12}$$

In applying the existence theorem we take these conditions as the equations F_0 referred to in § 8. Differentiating (47.12) and reducing by means of (46.7), we obtain a new set of conditions which together with (47.1) and (47.4) constitute the set F_1 of equations; and so on. Since the equations (47.12) are homogeneous and of third degree, the existence theorem assumes the more general form of § 8, and not that applying to the cases when all the equations are linear and homogeneous. If the coördinates are such that the components ξ^i are of the form (46.11), equations (47.12) reduce to $\Gamma^\alpha_{11} = 0$ $(\alpha = 2, \cdots, n)$. Combining this result with those of § 46, we have a means of defining the most general affine connection admitting a group G_1 of the type under discussion.

48. Continuous groups of collineations. If $\xi^i_{(\alpha)}$ for $\alpha = 1, \cdots, r$ determine infinitesimal collineations, we call $X_\alpha f \equiv \xi^i_{(\alpha)} \dfrac{\partial f}{\partial x^i}$ the generators of the collineations. Furthermore, we denote by $(X_\alpha, X_\beta) f$ the Poisson operator, that is,

$$(48.1)\quad (X_\alpha, X_\beta) f = \xi^i_{(\alpha)} \frac{\partial}{\partial x^i}\left(\xi^j_{(\beta)} \frac{\partial f}{\partial x^j}\right) - \xi^i_{(\beta)} \frac{\partial}{\partial x^i}\left(\xi^j_{(\alpha)} \frac{\partial f}{\partial x^j}\right).$$

We establish the following theorem:

If $X_\alpha f$ for $\alpha = 1, \cdots, r$ are the generators of infinitesimal collineations, so also are $(X_\alpha, X_\beta) f$ for $\alpha, \beta = 1, \cdots, r$ $(\alpha \neq \beta)$. Consider the case when $\alpha = 1$, $\beta = 2$. From (48.1) it follows that

$$(X_1, X_2) f = \xi^i \frac{\partial f}{\partial x^i},$$

where

$$(48.2)\quad \xi^i = \xi^h_{(1)} \frac{\partial \xi^i_{(2)}}{\partial x^h} - \xi^h_{(2)} \frac{\partial \xi^i_{(1)}}{\partial x^h} = \xi^h_{(1)} \xi^i_{(2),h} - \xi^h_{(2)} \xi^i_{(1),h}.$$

From these expressions and (46.7) we have in consequence of the identities (21.3) and (21.4),

$$\begin{aligned} \xi^i_{,j} = \xi^h_{(1),j} \xi^i_{(2),h} - \xi^h_{(2),j} \xi^i_{(1),h} + \xi^h_{(1)} \xi^l_{(2)} B^i_{jhl} \\ + \delta^i_j (\xi^h_{(1)} \varphi_{(2)h} - \xi^h_{(2)} \varphi_{(1)h}) + \xi^i_{(1)} \varphi_{(2)j} - \xi^i_{(2)} \varphi_{(1)j}. \end{aligned}$$

If we differentiate these equations covariantly with respect to x^k and in the reduction make use of (46.7) and (21.5), we obtain

$$\xi^i_{,jk} = \xi^h B^i_{jkh} + \delta^i_j \varphi_k + \delta^i_k \varphi_j,$$

where

$$\varphi_j = \xi^h_{(1),j} \varphi_{(2)h} - \xi^h_{(2),j} \varphi_{(1)h} + \xi^h_{(1)} \varphi_{(2)j,h} - \xi^h_{(2)} \varphi_{(1)j,h},$$

which establishes the theorem.

Suppose that a given space admits r independent infinitesimal collineations. From the above theorem and those of § 47 it follows that

$$(X_\alpha, X_\beta) f = c_{\alpha\beta}{}^\gamma X_\gamma f,$$

where the c's are constants. Hence as a consequence of the fundamental theorem of the theory of continuous groups* we have:

When, and only when, equations (46.7) *or* (46.9) *admit r independent solutions, the space admits an r parameter continuous group of projective or affine collineations.*†

49. Collineations in a Riemannian space. In order that an infinitesimal transformation (46.1) in a Riemannian space, with fundamental tensor g_{ij}, be a collineation, it is necessary that

$$(49.1) \qquad g'_{ij} = g_{ij}(x'^1, \cdots, x'^n) = g_{ij} + \frac{\partial g_{ij}}{\partial x^k} \xi^k \delta u$$

and that equations (46.5) be satisfied, when Γ^i_{jk} are replaced by the Christoffel symbols of the second kind formed with respect to g_{ij}. The latter conditions reduce, as in § 46, to

$$(49.2) \qquad \xi^h_{\ ,ij} = \xi^k R^h_{ijk} + \delta^h_j \varphi_{,i} + \delta^h_i \varphi_{,j},$$

where R^h_{ijk} are the components of the curvature tensor and $\varphi_{,i}$ are the components of a gradient; the latter follows from (47.2), since $S_{ij} = 0$ for a Riemannian space; in (49.2) covariant differentiation is with respect to the g's.

The quantities g'_{ij} given by (49.1) are the components in the x''s of a tensor whose components $\bar{g}_{ij}$ in the x's under the transformation (46.1) are given by

$$(49.3) \qquad \bar{g}_{ij} = g'_{kl} \frac{\partial x'^k}{\partial x^i} \frac{\partial x'^l}{\partial x^j} = g_{ij} + h_{ij}\, \delta u,$$

where

$$(49.4) \quad h_{ij} = \xi^k \frac{\partial g_{ij}}{\partial x^k} + g_{ik} \frac{\partial \xi^k}{\partial x^j} + g_{jk} \frac{\partial \xi^k}{\partial x^i} = \xi_{i,j} + \xi_{j,i},$$

and $\xi_i = g_{ij}\xi^j$. From (49.3) follow equations of the form

$$\left\{ {l \atop ij} \right\}_{\bar{g}} = \left\{ {l \atop ij} \right\}_g + a^l_{ij}\, \delta u,$$

* *Lie*, 1893, 1, p. 391; also *Bianchi*, 1918, 1, p. 97.

† Cf. *Knebelman*. 1927, 4.

where a subscript as in $\left\{ {l \atop ij} \right\}_{\bar{g}}$ indicates the form with respect to which the symbol is formed, a^l_{ij} being functions to be determined. Multiplying by $\bar{g}_{lk}$, summing for l and using the expressions (49.3) in the right-hand member of the equation, we have

$$[ij, k]_{\bar{g}} = [ij, k]_g + \left(h_{kl} \left\{ {l \atop ij} \right\}_g + g_{kl}\, a^l_{ij} \right) \delta u,$$

where $[ij, k]_{\bar{g}}$ are Christoffel symbols of the first kind formed with respect to $\bar{g}_{ij}$. In accordance with the definition of these symbols we have from (49.3)

$$[ij, k]_{\bar{g}} = [ij, k]_g + [ij, k]_h \,\delta u.$$

Consequently

$$[ij, k]_h = h_{kl} \left\{ {l \atop ij} \right\}_g + g_{kl}\, a^l_{ij}.$$

If we add to this equation the one obtained by interchanging i and k, the result may be written

$$h_{ik,j} = g_{kl}\, a^l_{ij} + g_{il}\, a^l_{kj}.$$

Substituting for $h_{ik,j}$ the expression from (49.4) and making use of (49.2) in the form

(49.5) $$\xi_{h,ij} = \xi^k R_{hijk} + g_{jh}\, \varphi_{,i} + g_{ih}\, \varphi_{,j};$$

we obtain

$$g_{kl} (a^l_{ij} - \delta^l_j \varphi_{,i} - \delta^l_i \varphi_{,j}) + g_{il} (a^l_{kj} - \delta^l_j \varphi_{,k} - \delta^l_k \varphi_{,j}) = 0.$$

When we add to this equation the one obtained from it by interchanging i and j and subtract the one obtained by interchanging j and k, we find that

$$a^l_{ij} = \delta^l_j \varphi_{,i} + \delta^l_i \varphi_{,j}.$$

Consequently we have

(49.6) $$\left\{ {l \atop ij} \right\}_{\bar{g}} = \left\{ {l \atop ij} \right\}_g + (\delta^l_j \varphi_{,i} + \delta^l_i \varphi_{,j}) \delta u$$

and

(49.7) $$h_{ij,k} = 2 g_{ij}\, \varphi_{,k} + g_{jk}\, \varphi_{,i} + g_{ik}\, \varphi_{,j}.^*$$

* For another method of obtaining equations (49.3) and (49.7) see 1926, 1, p. 228.

Each solution of (49.5) in which $\varphi_{,i} \neq 0$ determines a group G_1 of projective collineations of the Riemannian space. The results of § 48 can be applied to this case to determine whether a space admits a group G_r of projective collineations. From (49.5) we have

$$\xi_{h,ij} + \xi_{i,hj} = 2\, g_{ih}\, \varphi_{,j} + g_{jh}\, \varphi_{,i} + g_{ij}\, \varphi_{,h}.$$

From these equations we have:

A necessary and sufficient condition that a collineation of a Riemannian space be affine is that the first covariant derivative of $\xi_{h,i} + \xi_{i,h}$ *be zero.*

When in particular $\xi_{h,i} + \xi_{i,h} = 0$, then $\bar{g}_{ij} = g_{ij}$ and the collineation is a motion.*

* 1926, 1, p. 234.

CHAPTER IV

THE GEOMETRY OF SUB-SPACES

50. Covariant pseudonormal to a hypersurface. The vector-field ν^α. Consider a space V_{n+1} expressed in terms of coördinates y^α.* A *hypersurface* V_n is defined by an equation of the form

$$\varphi(y^1, \cdots, y^{n+1}) = 0, \tag{50.1}$$

where φ is irreducible. If we put

$$x^i = \varphi^i(y^1, \cdots, y^{n+1}), \qquad x^{n+1} = \varphi, \tag{50.2}$$

where the functions φ^i are arbitrary except that the jacobian of the $n+1$ φ's is different from zero, then equations (50.2) define a coördinate system x^α for which the given V_n is the hypersurface $x^{n+1} = 0$.

For any displacement in V_n at a point P of it we have

$$\frac{\partial \varphi}{\partial y^\alpha} dy^\alpha = 0, \tag{50.3}$$

and consequently the covariant vector ν_α at P, defined by

$$\nu_\alpha = \frac{\partial \varphi}{\partial y^\alpha} = \frac{\partial x^{n+1}}{\partial y^\alpha}, \tag{50.4}$$

is pseudo-orthogonal (§ 11) to every contravariant vector tangential to V_n at P. From (50.2) and (50.4) it is evident that the vector ν_α is independent of the choice of the functions φ^i in (50.2). We call it the *covariant pseudo-normal* to V_n.

* In this and the following sections greek indices take the values $1, \cdots, n+1$ and latin $1, \cdots, n$.

We define also a contravariant vector ν^α by the equations

$$\nu^\alpha = \frac{\partial y^\alpha}{\partial x^{n+1}}. \tag{50.5}$$

As thus defined ν^α are the components of the vector tangential to the curves of parameter x^{n+1}, that is, the curves along which all the x's except x^{n+1} are constant; we call them *transversals* of the hypersurface. Evidently ν^α depends upon the choice of the functions φ^i in (50.2). From (50.4) and (50.5) we have

$$\nu_\alpha \nu^\alpha = 1 \tag{50.6}$$

and

$$\nu^\alpha \frac{\partial x^i}{\partial y^\alpha} = 0. \tag{50.7}$$

If we change the curves of parameter x^{n+1}, we get a new vector of the type ν^α. Calling it $\bar{\nu}^\alpha$, we must have

$$\bar{\nu}^\alpha = \nu^\alpha + a^i \frac{\partial y^\alpha}{\partial x^i}, \tag{50.8}$$

if we require that $\bar{\nu}^\alpha \nu_\alpha = 1$.

Suppose that we have a set of functions $\psi^\alpha(y^1, \cdots, y^{n+1})$ such that $\psi^\alpha \frac{\partial \varphi}{\partial y^\alpha} = 1$, that is, $\psi^\alpha \nu_\alpha = 1$. If we put $\nu^\alpha = \psi^\alpha$, the condition (50.6) is satisfied. Moreover, if for the functions φ^i of (50.2) we take n independent solutions of the equations

$$\frac{dy^1}{\psi^1} = \cdots = \frac{dy^{n+1}}{\psi^{n+1}}, \tag{50.9}$$

then in the coördinates x^α the integral curves of (50.9) are the curves of parameter x^{n+1}. Thus for a given congruence ν^α not tangential to V_n we can define a coördinate system x^α satisfying the requirements of this section.

When equations (50.2) are solved for the y's, we have

$$y^\alpha = f^\alpha(x^1, \cdots, x^{n+1}) \tag{50.10}$$

or, as expansions in powers of x^{n+1},

$$(50.11)\quad y^\alpha = f_0^\alpha(x^1, \cdots, x^n) + f_1^\alpha(x^1, \cdots, x^n)x^{n+1} + \cdots.$$

Consequently V_n is defined by the parametric equations

$$(50.12)\qquad y^\alpha = f_0^\alpha(x^1, \cdots, x^n),$$

and at points of V_n the expressions for ν^α in the x's are

$$(50.13)\qquad (\nu^\alpha)_0 = f_1^\alpha(x^1, \cdots, x^n).$$

If in (50.11) we put

$$x^i = x'^i, \qquad x^{n+1} = x'^{n+1} F(x'^1, \cdots, x'^n),$$

then the curves of parameter x'^{n+1} are the same as those of parameter x^{n+1}, but in place of (50.13) we have

$$(\nu'^\alpha)_0 = f_1^\alpha(x'^1, \cdots, x'^n)F.$$

Thus we see in what manner the coördinate x^{n+1} may be changed, if the vector ν^α at points of V_n is not to change in direction.

In order to consider the effect of a change of the vector-field ν^α, we consider a transformation of coördinates of the form

$$(50.14)\qquad \begin{aligned} x'^i &= x^i + \psi_1^i(x^1, \cdots, x^n)x^{n+1} \\ &\qquad + \psi_2^i(x^1, \cdots, x^n)(x^{n+1})^2 + \cdots, \\ x'^{n+1} &= x^{n+1} + \psi_2^{n+1}(x^1, \cdots, x^n)(x^{n+1})^2 + \cdots, \end{aligned}$$

that is, a transformation such that in the two coördinate systems the space V_n is defined by $x^{n+1} = x'^{n+1} = 0$, and $x'^i = x^i$ at each point of V_n. From (50.5) it follows that in the x's the components of the field ν^α are

$$(50.15)\qquad \nu^i = 0, \qquad \nu^{n+1} = 1.$$

From (50.14) and (50.15) we have that the components of this field in the x''s are

$$
\begin{aligned}
\nu'^i &= \psi_1^i + 2\,\psi_2^i x^{n+1} + \cdots, \\
\nu'^{n+1} &= 1 + 2\,\psi_2^{n+1} x^{n+1} + \cdots.
\end{aligned} \tag{50.16}
$$

It follows at once that the curves of parameter x^{n+1} and of parameter x'^{n+1} have the same directions at each point of V_n, when, and only when,

$$
\psi_1^i (x^1, \cdots, x^n) = 0 \qquad (i = 1, \cdots, n). \tag{50.17}
$$

If we denote by y^α a general coördinate system, we have from (50.14)

$$
\begin{aligned}
\frac{\partial y^\alpha}{\partial x^i} &= \frac{\partial y^\alpha}{\partial x'^j}\left(\delta_i^j + \frac{\partial \psi_1^j}{\partial x^i} x^{n+1} + \cdots\right) \\
&\quad + \frac{\partial y^\alpha}{\partial x'^{n+1}}\left(\frac{\partial \psi_2^{n+1}}{\partial x^i}(x^{n+1})^2 + \cdots\right), \\
\frac{\partial y^\alpha}{\partial x^{n+1}} &= \frac{\partial y^\alpha}{\partial x'^j}(\psi_1^j + 2\,\psi_2^j x^{n+1} + \cdots) \\
&\quad + \frac{\partial y^\alpha}{\partial x'^{n+1}}(1 + 2\,\psi_2^{n+1} x^{n+1} + \cdots)
\end{aligned} \tag{50.18}
$$

The result (50.17) follows also from the last of these expressions. Also we have

$$
\begin{aligned}
\frac{\partial x'^i}{\partial y^\alpha} &= \frac{\partial x^j}{\partial y^\alpha}\left(\delta_j^i + \frac{\partial \psi_1^i}{\partial x^j} x^{n+1} + \cdots\right) \\
&\quad + \frac{\partial x^{n+1}}{\partial y^\alpha}(\psi_1^i + 2\,\psi_2^i x^{n+1} + \cdots), \\
\frac{\partial x'^{n+1}}{\partial y^\alpha} &= \frac{\partial x^j}{\partial y^\alpha}\left(\frac{\partial \psi_2^{n+1}}{\partial x^j}(x^{n+1})^2 + \cdots\right) \\
&\quad + \frac{\partial x^{n+1}}{\partial y^\alpha}(1 + 2\,\psi_2^{n+1} x^{n+1} + \cdots).
\end{aligned} \tag{50.19}
$$

From equations (50.18) and (50.19) we have that at points of V_n

$$
\begin{aligned}
\frac{\partial y^\alpha}{\partial x^i} &= \frac{\partial y^\alpha}{\partial x'^i}, \qquad & \frac{\partial y^\alpha}{\partial x^{n+1}} &= \frac{\partial y^\alpha}{\partial x'^j}\psi_1^j + \frac{\partial y^\alpha}{\partial x'^{n+1}}, \\
\frac{\partial x'^i}{\partial y^\alpha} &= \frac{\partial x^i}{\partial y^\alpha} + \frac{\partial x^{n+1}}{\partial y^\alpha}\psi_1^i, \qquad & \frac{\partial x'^{n+1}}{\partial y^\alpha} &= \frac{\partial x^{n+1}}{\partial y^\alpha}.
\end{aligned} \tag{50.20}
$$

Also from (50.18) we have that at points of V_n

$$\frac{\partial^2 y^\alpha}{\partial x^i \partial x^j} = \frac{\partial^2 y^\alpha}{\partial x'^i \partial x'^j},$$

$$\frac{\partial^2 y^\alpha}{\partial x^i \partial x^{n+1}} = \frac{\partial^2 y^\alpha}{\partial x'^i \partial x'^{n+1}} + \frac{\partial^2 y^\alpha}{\partial x'^i \partial x'^k} \psi_1^k + \frac{\partial y^\alpha}{\partial x'^j} \frac{\partial \psi_1^j}{\partial x'^i}. \tag{50.21}$$

If we write the last of equations (50.18) in the form

$$\nu^\alpha = \frac{\partial y^\alpha}{\partial x'^i} (\psi_1^i + 2 \psi_2^i x^{n+1} + \cdots) + \nu'^\alpha (1 + 2 \psi_2^{n+1} x^{n+1} + \cdots). \tag{50.22}$$

then at points of V_n

$$\frac{\partial \nu^\alpha}{\partial x^j} = \frac{\partial^2 y^\alpha}{\partial x^i \partial x^j} \psi_1^i + \frac{\partial y^\alpha}{\partial x^i} \frac{\partial \psi_1^i}{\partial x^j} + \frac{\partial \nu'^\alpha}{\partial x'^j}. \tag{50.23}$$

51. Transversals of a hypersurface which are paths of the enveloping space. We consider the transformation of coördinates in the space V_{n+1} defined by

$$\begin{aligned} y^\alpha = f_0^\alpha + f_1^\alpha x'^{n+1} - \frac{1}{2} (\overline{\Gamma}^\alpha_{\beta\gamma})_0 f_1^\beta f_1^\gamma (x'^{n+1})^2 \cdots \\ - \frac{1}{r!} (\overline{\Gamma}^\alpha_{\beta_1 \cdots \beta_r})_0 f_1^{\beta_1} \cdots f_1^{\beta_r} (x'^{n+1})^r + \cdots, \end{aligned} \tag{51.1}$$

where f_0^α and f_1^α are functions of $x'^1, \cdots, x'^n$. $\overline{\Gamma}^\alpha_{\beta\gamma}$ are the coefficients of the affine connection in the y's in V_{n+1}, $\overline{\Gamma}^\alpha_{\beta_1 \cdots \beta_r}$ are defined by (22.8), and the zero subscript indicates that in these functions y^α have been replaced by f_0^α. From the considerations of § 22 it follows that these expressions satisfy formally the equations

$$\frac{\partial^2 y^\alpha}{(\partial x'^{n+1})^2} + \overline{\Gamma}^\alpha_{\beta\gamma} \frac{\partial y^\beta}{\partial x'^{n+1}} \frac{\partial y^\gamma}{\partial x'^{n+1}} = 0.$$

Hence when constant values are given to $x'^1, \cdots, x'^n$, equations (51.1) define a path of V_{n+1}, the parameter being x'^{n+1}. The hypersurface $x'^{n+1} = 0$ is defined by

$$y^{\alpha} = f_0^{\alpha}(x'^1, \cdots, x'^n), \tag{51.2}$$

and the tangent vector ν^{α} at a point of this hypersurface to the path of the above set through the point has the components

$$(\nu^{\alpha})_0 = f_1^{\alpha}(x'^1, \cdots, x'^n). \tag{51.3}$$

When we effect upon the x''s a transformation of the form (50.14) in which $\psi_1^i = 0$, we get (50.11), in which the functions f_0^{α} and f_1^{α} are of the same form as in (51.1). Moreover, equations (51.2) and (51.3) are the same as (50.12) and (50.13) respectively. In what follows we shall use the equations in the form of § 50, when we are considering relations between the y's and x's, and the same notation with primes when the coördinates x'^{α} are under consideration. Thus in place of (50.2) we have

$$x'^i = \varphi'^i(y^1, \cdots, y^{n+1}), \quad x'^{n+1} = \varphi'(y^1, \cdots, y^{n+1}). \tag{51.4}$$

In particular, we remark that $\varphi' = 0$ is the equation of the hypersurface V_n. From the last of (50.14) it is seen that $\varphi' = \varphi F(y^1, \cdots, y^{n+1})$, where F is of such a form that $F = 1$ when $\varphi = 0$.

Differentiating the last of the equations (51.4) with respect to x'^{n+1}, we have

$$\frac{\partial \varphi'}{\partial y^{\alpha}} \nu'^{\alpha} = 1, \tag{51.5}$$

where, in consequence of (51.1)

$$\nu'^{\alpha} = f_1^{\alpha} - (\bar{\Gamma}^{\alpha}_{\beta\gamma})_0 f_1^{\beta} f_1^{\gamma} x'^{n+1} + \cdots. \tag{51.6}$$

Differentiating (51.5) with respect to x'^{n+1}, we have at points of V_n

$$(\nu'^{\alpha} \nu'^{\beta} \varphi'_{,\alpha\beta})_0 = 0. \tag{51.7}$$

Again differentiating (51.5) with respect to x'^i, we have

$$\left(\nu'^{\alpha} \frac{\partial^2 \varphi'}{\partial y^{\alpha} \partial y^{\beta}} \frac{\partial y^{\beta}}{\partial x'^i}\right)_0 + \left(\frac{\partial \varphi'}{\partial y^{\alpha}} \frac{\partial \nu'^{\alpha}}{\partial x'^i}\right)_0 = 0. \tag{51.8}$$

Proceeding in like manner with the last of equations (50.2), we have

$$(51.9)\qquad \left(\nu^\alpha \frac{\partial^2 \varphi}{\partial y^\alpha \partial y^\beta}\frac{\partial y^\beta}{\partial x^i}\right)_0 + \left(\frac{\partial \varphi}{\partial y^\alpha}\frac{\partial \nu^\alpha}{\partial x^i}\right)_0 = 0.$$

Since at points of V_n $(\nu'^\alpha)_0 = (\nu^\alpha)_0$ and $\left(\frac{\partial y^\alpha}{\partial x^i}\right)_0 = \left(\frac{\partial y^\alpha}{\partial x'^i}\right)_0$ [cf. (50.20)], we have from the equations (51.5) and

$$\nu^\alpha \frac{\partial \varphi}{\partial y^\alpha} = 1, \quad \frac{\partial \varphi}{\partial y^\alpha}\frac{\partial y^\alpha}{\partial x^i} = 0, \quad \frac{\partial \varphi'}{\partial y^\alpha}\frac{\partial y^\alpha}{\partial x'^i} = 0$$

that $\left(\frac{\partial \varphi}{\partial y^\alpha}\right)_0 = \left(\frac{\partial \varphi'}{\partial y^\alpha}\right)_0$. Consequently from (51.8) and (51.9), we have at points of V_n

$$(51.10)\qquad \nu^\alpha \frac{\partial y^\beta}{\partial x^i}(\varphi_{,\alpha\beta} - \varphi'_{,\alpha\beta}) = 0.$$

From this result and (50.3) it follows that

$$(51.11)\qquad (\nu^\alpha)_0 (\varphi_{,\alpha\beta} - \varphi'_{,\alpha\beta})_0 = (\nu_\beta)_0 f,$$

where, in consequence of (51.7) and (50.6)

$$(51.12)\qquad f_1^\alpha f_1^\beta (\varphi_{,\alpha\beta})_0 = f.$$

An application of these results is made in § 56.

52. Tensors in a hypersurface derived from tensors in the enveloping space. Let ξ_α and λ_α be the components in the y's and x's of a covariant vector-field in V_{n+1}; then

$$(52.1)\qquad \lambda_i = \xi_\alpha \frac{\partial y^\alpha}{\partial x^i}$$

and

$$(52.2)\qquad \lambda_{n+1} = \xi_\alpha \frac{\partial y^\alpha}{\partial x^{n+1}} = \xi_\alpha \nu^\alpha.$$

At points of V_n $(x^{n+1} = 0)$ the functions ξ_α are expressible as functions of $x^1, \cdots, x^n$, and for each value of α the quantities $\frac{\partial y^\alpha}{\partial x^i}$ are the components of a covariant vector in V_n. Hence λ_i given by (52.1) are the components of a covariant vector in V_n, which we say is *derived* from the given vector in V_{n+1}. In particular, the vector $\varphi \nu_\alpha$, where φ is an arbitrary function of the y's, is the most general covariant vector whose derived vector in V_n is a zero vector. Also we have at once that all vectors of the pencil $\xi_\alpha + \varrho \nu_\alpha$ have the same derived vector in V_n whatever be ϱ.

In like manner, if $a_{\alpha_1 \cdots \alpha_r}$ are the components in the y's of a tensor in V_{n+1}, then

$$(52.3) \qquad b_{i_1 \cdots i_r} = a_{\alpha_1 \cdots \alpha_r} \frac{\partial y^{\alpha_1}}{\partial x^{i_1}} \cdots \frac{\partial y^{\alpha_r}}{\partial x^{i_r}}$$

evaluated at points of V_n are the components of a tensor in V_n *derived* from the given tensor in V_{n+1}. From (52.1) and (52.3) it is evident that the tensor in V_n derived from a covariant tensor in the enveloping V_{n+1} is independent of the choice of the vector ν^α. This is readily seen also by observing that the quantities $b_{i_1 \cdots i_r}$ possess tensor character under transformations of the form (50.14) whatever be the functions ψ^i_1, as follows from (50.20). The same is true for the general transformations

$$(52.4) \quad x^i = \varphi^i(x'^1, \cdots, x'^{n+1}), \quad x^{n+1} = x'^{n+1} F(x'^1, \cdots, x'^{n+1}),$$

where F and its first derivatives are finite for $x'^{n+1} = 0$.

Let ξ^α and λ^α be the components in the y's and x's of a contravariant vector-field in V_{n+1}; then we have

$$(52.5) \qquad \lambda^i = \xi^\alpha \frac{\partial x^i}{\partial y^\alpha}$$

and

$$(52.6) \qquad \lambda^{n+1} = \xi^\alpha \nu_\alpha.$$

At points of V_n the functions ξ^α are expressible as functions of $x^1, \cdots, x^n$ and for each value of α the quantities $\frac{\partial x^i}{\partial y^\alpha}$

for $i = 1, \cdots, n$ are components of a contravariant vector in V_n under transformations of the form

$$x^i = \psi^i(x'^1, \cdots, x'^n), \quad x^{n+1} = x'^{n+1}. \tag{52.7}$$

We say, that λ^i defined by (52.5) is the vector in V_n *derived* from the given vector ξ^α in V_{n+1}.

A contravariant vector in V_n is necessarily one in V_{n+1}. As a vector in V_{n+1} the vector defined in V_n by (52.5) has the components

$$\bar{\lambda}^i = \xi^\alpha \frac{\partial x^i}{\partial y^\alpha}, \qquad \bar{\lambda}^{n+1} = 0. \tag{52.8}$$

Accordingly it follows from (52.6) that when a contravariant vector in V_{n+1} is tangential to V_n, it is identical with the derived vector in V_n. If it is not tangential to V_n and we denote by $\bar{\xi}^\alpha$ the components in the y's of the vector (52.8), we have

$$\bar{\xi}^\alpha = \bar{\lambda}^\beta \frac{\partial y^\alpha}{\partial x^\beta} = \bar{\lambda}^i \frac{\partial y^\alpha}{\partial x^i} = \xi^\beta \frac{\partial x^i}{\partial y^\beta} \frac{\partial y^\alpha}{\partial x^i} = \xi^\beta B^\alpha_\beta, \tag{52.9}$$

where

$$B^\alpha_\beta = \delta^\alpha_\beta - \nu^\alpha \nu_\beta. \tag{52.10}$$

Following Schouten* we say that the derived vector of a contravariant vector is the *tangential component* in V_n of the given vector. From the form of (52.9) and (52.10) it follows that unless the given vector is tangential to V_n its tangential component depends upon the choice of the vector ν^α. This may be seen also from (52.5). For, $\frac{\partial x^i}{\partial y^\alpha}$ being the cofactor of $\frac{\partial y^\alpha}{\partial x^i}$ in the determinant $\Delta = \left| \frac{\partial y^\alpha}{\partial x^i} \right|$ divided by Δ, it evidently depends upon ν^α.

When in (52.5) we repace ξ^α by ν^α, we find that the derived vector is a zero vector, and $\varphi \nu^\alpha$ is the only vector possessing this property. Consequently, when a vector ν^α has been chosen, in the system so defined its tangential component

* 1924, 1, p. 134.

being zero, this vector is analogous to the normal vector to a hypersurface in a Riemannian space in the sense that the tangential component is zero. This fact has led certain writers to refer to ν^α as a pseudonormal.* We do not use this term for a general choice of ν^α, because in a Riemannian geometry it would be confusing (cf. § 56).

In like manner, if $a^{\alpha_1 \cdots \alpha_r}$ are the components in the y's of a tensor in V_{n+1}, the quantities

$$(52.11) \qquad b^{i_1 \cdots i_r} = a^{\alpha_1 \cdots \alpha_r} \frac{\partial x^{i_1}}{\partial y^{\alpha_1}} \cdots \frac{\partial x^{i_r}}{\partial y^{\alpha_r}},$$

evaluated at points of V_n, are components of a tensor in V_n under transformations of the type (52.7), that is, general transformations in V_n but which in V_{n+1} do not change the vector ν^α. We call this tensor the *derived* tensor in V_n. The tensor in V_{n+1} with the components $b^{i_1 \cdots i_r}$ and $b^{\sigma_1 \cdots \sigma_r} = 0$, when one or more of the σ's is $n+1$, is called by Schouten the *tangential component* with respect to V_n. If $\bar{a}^{\alpha_1 \cdots \alpha_r}$ are its components in the y's, we have

$$(52.12) \qquad \bar{a}^{\alpha_1 \cdots \alpha_r} = a^{\beta_1 \cdots \beta_r} B^{\alpha_1}_{\beta_1} \cdots B^{\alpha_r}_{\beta_r},$$

where the B's are defined as in (52.10).

Since a covariant vector in V_n is equivalent to the bundle of contravariant vectors in V_n pseudoörthogonal to it (§ 11), it is evident that a covariant vector in V_n is not one in an enveloping V_{n+1}. If λ_α are the components in the x's of a covariant vector in V_{n+1}, the vector of components

$$(52.13) \qquad \bar{\lambda}_i = \lambda_i, \qquad \bar{\lambda}_{n+1} = 0$$

is equivalent to the bundle of contravariant vectors determined by ν^α and the contravariant vectors in V_n pseudoörthogonal to the derived vector of λ_α in V_n. In this sense the derived vector is the tangential component of the vector

* Cf. *Weyl*, 1922, 6, p. 154 and *Schouten*, 1924, 1, p. 134.

(52.13). If ξ_α are the components of the given vector in the y's, the components of (52.13) in the y's are

$$\overline{\xi}_\alpha = \xi_\beta B^\beta_\alpha.$$

In general, the derived tensor in V_n of a tensor $a_{\alpha_1 \cdots \alpha_r}$ in V_{n+1} is the tangential component of the tensor

$$\overline{a}_{\alpha_1 \cdots \alpha_r} = a_{\beta_1 \cdots \beta_r} B^{\beta_1}_{\alpha_1} \cdots B^{\beta_r}_{\alpha_r}. \tag{52.14}$$

Again if $a^{\alpha_1 \cdots \alpha_r}_{\beta_1 \cdots \beta_s}$ are the components in the y's of a mixed tensor in V_{n+1}, the quantities

$$b^{i_1 \cdots i_r}_{j_1 \cdots j_s} = a^{\alpha_1 \cdots \alpha_r}_{\beta_1 \cdots \beta_s} \frac{\partial x^{i_1}}{\partial y^{\alpha_1}} \cdots \frac{\partial x^{i_r}}{\partial y^{\alpha_r}} \frac{\partial y^{\beta_1}}{\partial x^{j_1}} \cdots \frac{\partial y^{\beta_s}}{\partial x^{j_s}} \tag{52.15}$$

evaluated at points of V_n are the components of a tensor under general transformations (52.7) and we call it the tensor *derived* from the given tensor in V_{n+1}. This derived tensor is the tangential component of the tensor whose components in the y's are

$$\overline{a}^{\alpha_1 \cdots \alpha_r}_{\beta_1 \cdots \beta_s} = a^{\gamma_1 \cdots \gamma_r}_{\delta_1 \cdots \delta_s} B^{\alpha_1}_{\gamma_1} \cdots B^{\alpha_r}_{\gamma_r} B^{\delta_1}_{\beta_1} \cdots B^{\delta_s}_{\beta_s}. \tag{52.16}$$

We call each of the tensors defined by (52.12), (52.14) and (52.16) the *associate* of the given tensor with respect to V_n.

From (52.10), (50.3), (50.6) and (50.7) we have

$$\begin{aligned} B^\beta_\alpha \frac{\partial x^i}{\partial y^\beta} &= \frac{\partial x^i}{\partial y^\alpha}, & B^\beta_\alpha \frac{\partial x^{n+1}}{\partial y^\beta} &= 0, \\ B^\alpha_\beta \frac{\partial y^\beta}{\partial x^i} &= \frac{\partial y^\alpha}{\partial x^i}, & B^\alpha_\beta \frac{\partial y^\beta}{\partial x^{n+1}} &= 0. \end{aligned} \tag{52.17}$$

Because of these identities we have from (52.15)

$$b^{i_1 \cdots i_r}_{j_1 \cdots j_s} = \overline{a}^{\alpha_1 \cdots \alpha_r}_{\beta_1 \cdots \beta_s} \frac{\partial x^{i_1}}{\partial y^{\alpha_1}} \cdots \frac{\partial y^{\beta_s}}{\partial x^{j_s}}. \tag{52.18}$$

Similar results follow from (52.3) and (52.11). Hence we have:

The derived tensors in V_n of a given tensor in V_{n+1} and of its associate with respect to V_n are equivalent.

53. Symmetric connection induced in a hypersurface. If $\Gamma^\alpha_{\beta\gamma}$ and $\overline{\Gamma}^\alpha_{\beta\gamma}$ are the coefficients of a symmetric connection in a V_{n+1} in coördinates x^α and y^α respectively, we have from equations of the form (5.6)

$$(53.1) \qquad \Gamma^i_{jk} = \left(\frac{\partial^2 y^\alpha}{\partial x^j\, \partial x^k} + \overline{\Gamma}^\alpha_{\beta\gamma} \frac{\partial y^\beta}{\partial x^j} \frac{\partial y^\gamma}{\partial x^k}\right) \frac{\partial x^i}{\partial y^\alpha}.$$

At points of V_n ($x^{n+1} = 0$) the values of Γ^i_{jk} depend upon the choice of the vector ν^α as is evident from (53.1). However if $\Gamma'^\alpha_{\beta\gamma}$ are the coefficients for V_{n+1} in the x''s defined by (52.7), from equations analogous to (53.1) we obtain

$$(53.2) \qquad \Gamma^i_{jk} = \left(\frac{\partial^2 x'^l}{\partial x^j\, \partial x^k} + \Gamma'^l_{pq} \frac{\partial x'^p}{\partial x^j} \frac{\partial x'^q}{\partial x^k}\right) \frac{\partial x^i}{\partial x'^l}.$$

Consequently Γ^i_{jk} and Γ'^i_{jk} are the coefficients of the same connection in V_n; we call it the connection *induced* in V_n by that in V_{n+1} for a *given choice of the vector* ν^α. This qualification is seen to be necessary from (53.2); for, in case of transformations of the type (52.4) there are in the last member of (53.2) the added terms $\Gamma'^{n+1}_{pq} \dfrac{\partial x'^p}{\partial x^j} \dfrac{\partial x'^q}{\partial x^k} \dfrac{\partial x^i}{\partial x'^{n+1}}$.*

For an asymmetric connection in V_{n+1}, equations (53.1) hold and the skew-symmetric part of the induced connection is given by

$$(53.3) \qquad \Omega^i_{jk} = \overline{\Omega}^\alpha_{\beta\gamma} \frac{\partial y^\beta}{\partial x^j} \frac{\partial y^\gamma}{\partial x^k} \frac{\partial x^i}{\partial y^\alpha},$$

that is, the tensor Ω^i_{jk} of the induced connection is the derived tensor of the tensor $\overline{\Omega}^\alpha_{\beta\gamma}$ of the connection in V_{n+1}.

Consider, in particular, the case when the affine connection in V_{n+1} is that of a Riemannian space, determined by the

* In § 56 we obtain the relations between the coefficients of two induced connections for different choices of ν^α.

fundamental tensor of components $a_{\alpha\beta}$ and $g_{\alpha\beta}$ in the y's and x's respectively. Then we have

$$(53.4) \qquad a_{\alpha\beta}\frac{\partial y^\alpha}{\partial x^j}\frac{\partial y^\beta}{\partial x^k} = g_{jk}.$$

From these equations we have* at points of V_n $(x^{n+1} = 0)$

$$(53.5) \quad a_{\alpha\beta}\frac{\partial y^\beta}{\partial x^k}\left(\frac{\partial^2 y^\alpha}{\partial x^i \partial x^j} - \frac{\partial y^\alpha}{\partial x^h}\begin{Bmatrix} h \\ ij \end{Bmatrix}_g + \begin{Bmatrix} \alpha \\ \mu\nu \end{Bmatrix}_a \frac{\partial y^\mu}{\partial x^i}\frac{\partial y^\nu}{\partial x^j}\right) = 0,$$

where $\begin{Bmatrix} \alpha \\ \mu\nu \end{Bmatrix}_a$ are formed with respect to the fundamental form $a_{\alpha\beta}$ in V_{n+1} and $\begin{Bmatrix} h \\ ij \end{Bmatrix}_g$ with respect to the fundamental form g_{ij} in V_n. From equations of the form (5.6) we have

$$\frac{\partial^2 y^\alpha}{\partial x^i \partial x^j} + \begin{Bmatrix} \alpha \\ \mu\nu \end{Bmatrix}_a \frac{\partial y^\mu}{\partial x^i}\frac{\partial y^\nu}{\partial x^j} = \Gamma^h_{ij}\frac{\partial y^\alpha}{\partial x^h} + \Gamma^{n+1}_{ij}\frac{\partial y^\alpha}{\partial x^{n+1}}.$$

Substituting in the preceding equations, we obtain

$$\left(\Gamma^h_{ij} - \begin{Bmatrix} h \\ ij \end{Bmatrix}_g\right) g_{hk} + g_{k\,n+1}\Gamma^{n+1}_{ij} = 0.$$

In order that $\Gamma^h_{ij} = \begin{Bmatrix} h \\ ij \end{Bmatrix}_g$, it is necessary and sufficient that $g_{k\,n+1} = 0$ or $\Gamma^{n+1}_{ij} = 0$. In the former case the curves of parameter x^{n+1} meet V_n orthogonally. In the latter case we have

$$\frac{\partial^2 y^\alpha}{\partial x^i \partial x^j} - \begin{Bmatrix} h \\ ij \end{Bmatrix}_g \frac{\partial y^\alpha}{\partial x^h} + \begin{Bmatrix} \alpha \\ \mu\nu \end{Bmatrix}_a \frac{\partial y^\mu}{\partial x^i}\frac{\partial y^\nu}{\partial x^j} = 0,$$

from which it follows that V_n is a totally-geodesic hypersurface of V_{n+1}.† Hence we have:

A necessary and sufficient condition that the coefficients of the induced connection in a hypersurface of a space with a Riemannian connection be the Christoffel symbols of the second kind formed with respect to the derived fundamental

* Cf., 1926, 1, p. 147.

† Cf., 1926, 1, equations (43.4) p. 147 and (54.1) p. 183.

tensor is that the vector ν^α *be orthogonal to the hypersurface, or that the latter be totally-geodesic.*

We indicate by one or more subscripts preceded by a semi-colon covariant differentiation with respect to the induced connection.

From the definition (§ 52) of the associate of a given tensor in V_{n+1} with respect to V_n it follows that the components in the x's of the associate tensor for which one or more of the indices are $n+1$ are zero. Consequently from the general law of components of a tensor in two coördinate systems we have from (52.18)

$$(53.6)\quad b^{i_1\cdots i_r}_{j_1\cdots j_s;k} = \bar{a}^{\alpha_1\cdots\alpha_r}_{\beta_1\cdots\beta_s,\gamma}\frac{\partial x^{i_1}}{\partial y^{\alpha_1}}\cdots\frac{\partial x^{i_r}}{\partial y^{\alpha_r}}\frac{\partial y^{\beta_1}}{\partial x^{j_1}}\cdots\frac{\partial y^{\beta_s}}{\partial x^{j_s}}\frac{\partial y^\gamma}{\partial x^k}.$$

Hence we have:

The first covariant derivative in a hypersurface of the derived tensor of a given tensor in the enveloping space V_{n+1} *is the derived tensor of the covariant derivative in* V_{n+1} *of the associate of the given tensor with respect to the hypersurface.*

It should be remarked that although the derived tensor of a covariant tensor is independent of the choice of ν^α, its covariant derivative in the hypersurface as a derived tensor does depend upon ν^α.*

Equations (53.6) do not hold for derivatives of higher order. However, the second covariant derivative of $b^{i_1\cdots i_r}_{j_1\cdots j_s}$ is the derived tensor of $\left(\bar{a}^{\alpha_1\cdots\alpha_r}_{\beta_1\cdots\beta_s,\gamma}B^\gamma_\delta\right)_{,\sigma}$. By continuing this process we obtain derivatives of any order (cf. § 54).

54. Fundamental derived tensors in a hypersurface. We denote by ω_{ij} the tensor in V_n derived from the tensor $\nu_{\alpha,\beta}$ in V_{n+1}, that is,

$$(54.1)\qquad \omega_{ij} = \nu_{\alpha,\beta}\frac{\partial y^\alpha}{\partial x^i}\frac{\partial y^\beta}{\partial x^j}.$$

From the form of these expressions it is evident that this tensor is independent of the choice of ν^α. In the x's the components of the vector ν_α are

* Cf. *Schouten*, 1924, 1, p. 137.

$$\nu_i = 0, \quad \nu_{n+1} = 1, \tag{54.2}$$

as follows from (50.4). Consequently in the x's equations (54.1) are

$$\omega_{ij} = -\Gamma^{n+1}_{ij}. \tag{54.3}$$

Evidently ω_{ij} is symmetric.

For a given choice of ν^α we have the tensor $\nu^\alpha_{,\beta}$ and the vector $\nu^\alpha_{,\beta}\nu_\alpha$. We denote by l^i_j and l_i the components of their derived tensors in V_n, that is,

$$l^i_j = \nu^\alpha_{,\beta} \frac{\partial x^i}{\partial y^\alpha} \frac{\partial y^\beta}{\partial x^j} \tag{54.4}$$

and

$$l_i = \nu^\alpha_{,\beta}\nu_\alpha \frac{\partial y^\beta}{\partial x^i}. \tag{54.5}$$

In the x's the components of the vector ν^α are given by (50.15) and consequently in the x's we have

$$l^i_j = \Gamma^i_{n+1\,j} \tag{54.6}$$

and

$$l_i = \Gamma^{n+1}_{n+1\,i}. \tag{54.7}$$

Because of (50.6) equations (54.5) can be written as

$$l_i = -\nu_{\alpha,\beta}\nu^\alpha \frac{\partial y^\beta}{\partial x^i}. \tag{54.8}$$

With the aid of the tensor ω_{ij} we are able to express the covariant derivative of a derived tensor in terms of the derived tensor of the covariant derivative of the given tensor and other derived tensors. Thus if b_{ij} are the components of the derived tensor of $a_{\alpha\beta}$, we have from (53.6)

$$b_{ij;k} = (a_{\alpha\beta} B^\alpha_\gamma B^\beta_\delta)_{,\sigma} \frac{\partial y^\gamma}{\partial x^i} \frac{\partial y^\delta}{\partial x^j} \frac{\partial y^\sigma}{\partial x^k}.$$

With the aid of (52.17) we obtain

$$b_{ij;k} = a_{\alpha\beta,\sigma} \frac{\partial y^\alpha}{\partial x^i} \frac{\partial y^\beta}{\partial x^j} \frac{\partial y^\sigma}{\partial x^k} - a_{\alpha\beta}\left(\nu^\alpha \frac{\partial y^\beta}{\partial x^j}\omega_{ik} + \nu^\beta \frac{\partial y^\alpha}{\partial x^i}\omega_{jk}\right).$$

55. The generalized equations of Gauss and Codazzi. If $\Gamma^\alpha_{\beta\gamma}$ and $\bar{\Gamma}^\alpha_{\beta\gamma}$ are the coefficients of an affine connection in a V_{n+1} in coördinates x^α and y^α respectively, we have from (5.6)

$$\frac{\partial^2 y^\alpha}{\partial x^i \partial x^j} + \bar{\Gamma}^\alpha_{\beta\gamma} \frac{\partial y^\beta}{\partial x^i} \frac{\partial y^\gamma}{\partial x^j} = \Gamma^\beta_{ij} \frac{\partial y^\alpha}{\partial x^\beta}. \tag{55.1}$$

At points of the V_n ($x^{n+1} = 0$) these equations can be written, in consequence of (54.3), in the form

$$y^\alpha_{;ij} + \bar{\Gamma}^\alpha_{\beta\gamma} \frac{\partial y^\beta}{\partial x^i} \frac{\partial y^\gamma}{\partial x^j} = -\omega_{ij} \nu^\alpha, \tag{55.2}$$

the first term being the second covariant derivative of y^α with respect to the induced connection in V_n.

When in equations (55.1) we replace i by $n+1$, we have

$$\frac{\partial \nu^\alpha}{\partial x^j} + \bar{\Gamma}^\alpha_{\beta\gamma} \nu^\beta \frac{\partial y^\gamma}{\partial x^j} = \Gamma^\beta_{n+1\,j} \frac{\partial y^\alpha}{\partial x^\beta}.$$

At points of V_n these equations become, in consequence of (54.6) and (54.7),

$$\frac{\partial \nu^\alpha}{\partial x^j} + \bar{\Gamma}^\alpha_{\beta\gamma} \nu^\beta \frac{\partial y^\gamma}{\partial x^j} = l^i_j \frac{\partial y^\alpha}{\partial x^i} + l_j \nu^\alpha. \tag{55.3}$$

We desire to find the conditions which the tensors ω_{ij}, l^i_j and l_i must satisfy in order that equations (55.2) and (55.3) be consistent. The conditions of integrability of (55.2) are of the form [cf. (6.4)]

$$y^\alpha_{;ijk} - y^\alpha_{;ikj} = \frac{\partial y^\alpha}{\partial x^m} B^m_{ijk}, \tag{55.4}$$

where B^m_{ijk} is the curvature tensor of V_n formed with respect to Γ^i_{jk}.

When the expressions (55.2) are substituted in (55.4), the resulting equations are reducible by means of (55.2) and (55.3) to

$$\begin{aligned} \frac{\partial y^\alpha}{\partial x^m} B^m_{ijk} = \bar{B}^\alpha_{\beta\gamma\delta} \frac{\partial y^\beta}{\partial x^i} \frac{\partial y^\gamma}{\partial x^j} \frac{\partial y^\delta}{\partial x^k} + (\omega_{ik} l^m_j - \omega_{ij} l^m_k) \frac{\partial y^\alpha}{\partial x^m} \\ - \nu^\alpha (\omega_{ij;k} - \omega_{ik;j} + \omega_{ij} l_k - \omega_{ik} l_j), \end{aligned} \tag{55.5}$$

where $\bar{B}^{\alpha}_{\beta\gamma\delta}$ is the curvature tensor of V_{n+1} in the y's evaluated at points of V_n. If these equations be multiplied by $\frac{\partial x^h}{\partial y^\alpha}$ and by ν_α and α be summed, we have respectively, in consequence of (50.3), (50.6) and (50.7),

$$(55.6)\quad B^h_{ijk} = \bar{B}^{\alpha}_{\beta\gamma\delta}\frac{\partial y^\beta}{\partial x^i}\frac{\partial y^\gamma}{\partial x^j}\frac{\partial y^\delta}{\partial x^k}\frac{\partial x^h}{\partial y^\alpha} + \omega_{ik}\, l^h_j - \omega_{ij}\, l^h_k,$$

$$(55.7)\quad \omega_{ij;k} - \omega_{ik;j} = \bar{B}^{\alpha}_{\beta\gamma\delta}\frac{\partial y^\beta}{\partial x^i}\frac{\partial y^\gamma}{\partial x^j}\frac{\partial y^\delta}{\partial x^k}\nu_\alpha + \omega_{ik}\, l_j - \omega_{ij}\, l_k.$$

The conditions of integrability of (55.3) are [cf. (6.3)]

$$\nu^\alpha_{,ij} - \nu^\alpha_{,ji} = 0.$$

Substituting from (55.3) in these equations and proceeding as above, we obtain

$$(55.8)\quad l^h_{i;j} - l^h_{j;i} + l_i\, l^h_j - l_j\, l^h_i + \bar{B}^{\alpha}_{\beta\gamma\delta}\,\nu^\beta\frac{\partial y^\gamma}{\partial x^i}\frac{\partial y^\delta}{\partial x^j}\frac{\partial x^h}{\partial y^\alpha} = 0,$$

$$(55.9)\quad l_{i;j} - l_{j;i} + l^h_j\,\omega_{hi} - l^h_i\,\omega_{hj} + \bar{B}^{\alpha}_{\beta\gamma\delta}\,\nu_\alpha\,\nu^\beta\frac{\partial y^\gamma}{\partial x^i}\frac{\partial y^\delta}{\partial x^j} = 0.$$

Equations (55.2) and (55.3) are generalizations of well-known equations in the theory of surfaces in euclidean space,* and of equations in a general Riemannian space,† ω_{ij} being the components of the second fundamental tensor and ν^α the unit vector normal to V_n. In these cases the processes followed above lead to the Gauss and Codazzi equations. Equations (55.7) and (55.8) are equivalent in these cases and (55.9) are satisfied identically.

Accordingly we call (55.6) the *generalized equations of Gauss* and (55.7) and (55.8) the *generalized equations of Codazzi.*‡

* 1909, 1, p. 154.

† 1926, 1, pp. 147, 148.

‡ Cf. *Schouten*, 1924, 1, p. 140.

From the equations of Gauss it is seen that the curvature tensor of the induced connection is not ordinarily the derived tensor of the curvature tensor of the enveloping space.

When equations (55.6) are contracted for h and i, we obtain

$$(55.10)\quad B^i_{ijk} = \bar{B}^\alpha_{\beta\gamma\delta}\,(\delta^\beta_\alpha - \nu_\alpha\,\nu^\beta)\frac{\partial y^\gamma}{\partial x^j}\frac{\partial y^\delta}{\partial x^k} + \omega_{ik}\,l^i_j - \omega_{ij}\,l^i_k.$$

By means of (55.9) and the theorem of § 5 these equations are equivalent to

$$(55.11)\quad B^i_{ijk} = \left(\frac{\partial \bar{a}_\delta}{\partial y^\gamma} - \frac{\partial \bar{a}_\gamma}{\partial y^\delta}\right)\frac{\partial y^\gamma}{\partial x^j}\frac{\partial y^\delta}{\partial x^k} + \frac{\partial l_j}{\partial x^k} - \frac{\partial l_k}{\partial x^j}.$$

When the connection is asymmetric, the results of the preceding sections are essentially unaltered and the corresponding equations are obtained on replacing Γ^i_{jk} by L^i_{jk}; then the tensor ω_{ij} is not symmetric in i and j. The generalized equations of Gauss for this case are obtained from (55.6) on replacing B^h_{ijk} and $\bar{B}^\alpha_{\beta\gamma\delta}$ by L^h_{ijk} and $\bar{L}^\alpha_{\beta\gamma\delta}$. In the right-hand member of (55.7) there is the added term $-2\,\Omega^h_{jk}\,\omega_{ih}$; and in the left-hand members of (55.8) and (55.9) the added terms $2\,\Omega^k_{ij}\,l^h_k$ and $2\,l_h\,\Omega^h_{ij}$ respectively.

56. Contravariant pseudonormal. We consider the effect upon the coefficients of the induced linear conection and upon the derived tensors l^i_j and l_i of a change in the vectors ν^α at points of V_n, and to this end make use of the transformation (50.14). For the coördinate system x'^i the coefficients of the induced connection in V_n are given by

$$(56.1)\quad \Gamma'^i_{jk} = \left(\frac{\partial^2 y^\alpha}{\partial x'^j\,\partial x'^k} + \bar{\Gamma}^\alpha_{\beta\gamma}\frac{\partial y^\beta}{\partial x'^j}\frac{\partial y^\gamma}{\partial x'^k}\right)\frac{\partial x'^i}{\partial y^\alpha}.$$

In consequence of (50.20) and (50.21), it follows from the above equations, (53.1) and (54.3) that at points of V_n

$$(56.2)\quad \Gamma'^i_{jk} = \Gamma^i_{jk} - \omega_{jk}\,\psi^i_1.$$

If B'^h_{ijk} denote the components of the curvature tensor in Γ'^i_{jk}, we have

$$B'^{h}_{ijk} = B^{h}_{ijk} + \psi^h_1 [\omega_{ij;k} - \omega_{ik;j} + \psi^l_1 (\omega_{ik}\, \omega_{lj} - \omega_{ij}\, \omega_{lk})] + \omega_{ij}\, \psi_1{}^h{}_{;k} - \omega_{ik}\, \psi_1{}^h{}_{;j}. \tag{56.3}$$

From (50.22) and (50.23) we have at points of V_n

$$\begin{aligned} \nu'^{\alpha}{}_{,\beta} \frac{\partial y^\beta}{\partial x^j} &= \nu^{\alpha}{}_{,\beta} \frac{\partial y^\beta}{\partial x^j} - \frac{\partial y^\alpha}{\partial x^i} \frac{\partial \psi^i_1}{\partial x^j} \\ &\quad - \left(\frac{\partial^2 y^\alpha}{\partial x^i \partial x^j} + \overline{\Gamma}^\alpha_{\beta\gamma} \frac{\partial y^\beta}{\partial x^i} \frac{\partial y^\gamma}{\partial x^j} \right) \psi^i_1 \\ &= \nu^{\alpha}{}_{,\beta} \frac{\partial y^\beta}{\partial x^j} - \psi_1{}^i{}_{;j} \frac{\partial y^\alpha}{\partial x^i} + \omega_{ij}\, \psi^i_1\, \nu^\alpha. \end{aligned} \tag{56.4}$$

In consequence of these equations, (50.20) and the definition (54.4) of the tensor l^i_j we have

$$l'^i_j = l^i_j - \psi_1{}^i{}_{;j} + \psi^i_1 (l_j + \omega_{kj}\, \psi^k_1). \tag{56.5}$$

Since $\nu'_\alpha = \nu_\alpha$ at points of V_n, as follows from (50.20), we have in like manner from (54.5)

$$l'_i = l_i + \omega_{ij}\, \psi^j_1. \tag{56.6}$$

It is readily found that equations (55.6), (55.7), (55.8) and (55.9) and similar ones for the induced connection Γ'^i_{jk} are consistent in view of the above relations.

As an immediate consequence of (56.6) we have:

When the determinant $|\omega_{ij}|$ *is not zero, a vector-field* ν^α *at points of* V_n *is uniquely determined with respect to which the vector* l_i *is zero.**

Also we have:

When the determinant $|\omega_{ij}|$ *is of rank* $n - r$ $(r > 0)$, *a vector-field* ν^α *at points of* V_n *with respect to which the vector* l_i *is zero cannot be obtained unless the matrix*

$$\left\| \begin{matrix} \omega_{11} & \cdots & \omega_{1n} & l_1 \\ \cdot & \cdot & \cdot & \cdot \\ \cdot & \cdot & \cdot & \cdot \\ \omega_{n1} & \cdots & \omega_{nn} & l_n \end{matrix} \right\|$$

* This theorem is due to *Schouten*, 1924, 1, p. 143.

is of rank $n-r$; in this case the determination of ν^α involves r arbitrary functions.

It is seen from (54.8) that the vector l_i vanishes at all points of V_n, when, and only when,

$$\nu^\alpha \nu_{\alpha,\beta} = f\nu_\beta \tag{56.7}$$

at points of V_n, where f is a function of the y's which may be zero. These equations are, in consequence of (50.4)

$$\nu^\alpha \varphi_{,\alpha\beta} = f\varphi_{,\beta}, \tag{56.8}$$

from which it follows that

$$\nu^\alpha \nu^\beta \varphi_{,\alpha\beta} = f.$$

If the determinant $|\varphi_{,\alpha\beta}|$ is of rank $n+1$, by a suitable choice of f a unique vector ν^α is given by (56.8) so that $\nu^\alpha \dfrac{\partial \varphi}{\partial x^\alpha} = 1$. If we retain this field at points of V_n, and apply the method of § 51 to obtain a coördinate system x'^α so that the curves of parameter x'^{n+1} are paths, whose tangents at points of V_n have the direction of ν^α, it follows from (51.11) and (51.12) that $\nu^\alpha \varphi'_{,\alpha\beta} = 0$. This is a generalization of the situation in a Riemannian space when a family of hypersurfaces $\psi =$ const. are geodesically parallel, the function ψ being chosen so that*

$$g^{\alpha\beta} \psi_{,\alpha} \psi_{,\beta} = 1.$$

From this we have $g^{\alpha\beta} \psi_{,\alpha} \psi_{,\beta\gamma} = 0$. Hence the vector ν^α in this case is the normal to the hypersurface. Accordingly when there is a field ν^α satisfying the conditions of the first theorem, we say that the vectors ν^α are the *contravariant pseudonormals* to the hypersurface.

When for a given choice of the field ν^α the vector l_i is a gradient, and we put

$$l_i = -\frac{\partial \log \theta}{\partial x^i} = -\frac{\partial \log \theta}{\partial y^\alpha} \frac{\partial y^\alpha}{\partial x^i},$$

* 1926, 1, p. 57.

we have from (54.5) at points of V_n

$$(\theta \nu^\alpha)_{,\beta} \nu_\alpha \frac{\partial y^\beta}{\partial x^i} = 0.$$

Hence, if in accordance with the results of § 50 we take the field $\theta \nu^\alpha$ at points of V_n, the corresponding vector l_i is zero. Accordingly we have:

When for a given choice of the field ν^α, the vector l_i is a gradient, there exists a function θ such that for the field $\theta \nu^\alpha$ the vector l_i is zero.

Thus the case where l_i is zero is equivalent to that where it is a gradient so far as the direction of the field ν^α goes.

If ψ^j_1 in (56.6) are chosen so that $l'_i = 0$, from (55.11) we have

$$B'^i_{ijk} = B^i_{ijk} - \frac{\partial l_j}{\partial x^k} + \frac{\partial l_k}{\partial x^j} = \overline{B}^\alpha_{\alpha\gamma\delta} \frac{\partial y^\gamma}{\partial x^j} \frac{\partial y^\delta}{\partial x^k}.$$

Consequently if $\overline{B}^\alpha_{\alpha\gamma\delta} = 0$, we have $B'^i_{ijk} = 0$. Conversely, if $B^i_{ijk} = \overline{B}^\alpha_{\alpha\beta\gamma} = 0$, it follows from (55.11) that l_i is a gradient. Hence we have:

*When for an affinely connected space $\overline{B}^\alpha_{\alpha\beta\gamma} = 0$, a necessary and sufficient condition that the directions of a vector-field ν^α at points of a hypersurface be such that l_i be a zero vector is that the corresponding tensor B^i_{ijk} be zero.**

When equations (55.3) are written in the form

$$\nu^\alpha{}_{,\beta} \frac{\partial y^\beta}{\partial x^j} = l^i_j \frac{\partial y^\alpha}{\partial x^i} + l_j \nu^\alpha,$$

the quantities on the left are the components of the associate direction of the vector ν^α for a displacement in the direction of a curve of parameter x^j in V_n. Hence we have:

When a hypersurface admits a contravariant pseudonormal, the associate direction of this normal in the enveloping space for any displacement in the hypersurface is tangential to the hypersurface.

* This theorem has been established by *Schouten*, 1924, 1, p. 142.

This property of the contravariant pseudonormal is possessed also by the normal to a hypersurface in a Riemannian space,* which justifies further the term pseudonormal.

57. Fundamental equations when the determinant ω is not zero. In this section we consider the case when the determinant ω is not zero at all points of V_n and we understand that the unique vector ν^α has been chosen for which $l_i = 0$. In this case a tensor g^{ij} is defined by the equations

$$g^{ij}\,\omega_{jk} = l^i_k. \tag{57.1}$$

We assume that the determinant

$$l \equiv |l^i_k| \tag{57.2}$$

is not zero.† Then it follows from (57.1) that the determinant $|g^{ij}|$ is not zero, and a tensor g_{ij} is uniquely defined by the equations

$$g_{ij}\,g^{jk} = \delta^k_i, \qquad g^{ij}\,g_{jk} = \delta^i_k. \tag{57.3}$$

From these equations and (57.1) we have

$$\omega_{ij} = g_{ik}\,l^k_j. \tag{57.4}$$

If we put

$$a_{\alpha\beta} = g_{ij}\frac{\partial x^i}{\partial y^\alpha}\frac{\partial x^j}{\partial y^\beta} + \nu_\alpha\,\nu_\beta, \tag{57.5}$$

it is evident from the form of these expressions that $a_{\alpha\beta}$ are the components in the y's of a tensor in V_{n+1}. From (57.5) we have, in consequence of (50.4), (50.6) and (50.7),

$$a_{\alpha\beta}\frac{\partial y^\alpha}{\partial x^i}\frac{\partial y^\beta}{\partial x^j} = g_{ij}, \tag{57.6}$$

$$a_{\alpha\beta}\frac{\partial y^\alpha}{\partial x^i}\,\nu^\beta = a_{\alpha\beta}\,\nu^\alpha\frac{\partial y^\beta}{\partial x^i} = 0 \tag{57.7}$$

* 1926, 1, p. 148.

† This is an additional assumption. For it follows from (56.5) that for a transformation (50.14) preserving ν^α (i. e., $\psi^i_1 = 0$) l^i_j are unaltered.

and

(57.8) $$a_{\alpha\beta}\,\nu^\alpha\,\nu^\beta = 1.$$

From these equations it is seen that for the determinants

(57.9) $$a \equiv |a_{\alpha\beta}|, \quad g \equiv |g_{ij}|$$

we have

(57.10) $$a\left|\frac{\partial y^\alpha}{\partial x^\beta}\right|^2 = g.$$

Consequently $a \neq 0$ in the case under consideration. Accordingly a tensor $a^{\alpha\beta}$ is defined by

(57.11) $$a_{\alpha\beta}\,a^{\beta\gamma} = \delta^\gamma_\alpha, \quad a^{\gamma\beta}\,a_{\beta\alpha} = \delta^\gamma_\alpha.$$

As a consequence of (57.7), (57.8), (50.3) and (50.6) we have

(57.12) $$a_{\alpha\beta}\,\nu^\alpha = a_{\beta\alpha}\,\nu^\alpha = \nu_\beta.$$

Moreover from (57.11) and (57.12) we have

(57.13) $$\nu^\alpha = \nu_\beta\,a^{\beta\alpha} = \nu_\beta\,a^{\alpha\beta}.$$

These equations are a generalization of the equations of a Riemannian space connecting the contravariant and covariant components of the normal to V_n.

If equations (55.6) are multiplied by g_{rh} and summed for h, we obtain, in consequence of (57.12),

(57.14) $$B_{rijk} = \overline{B}_{\alpha\beta\gamma\delta}\frac{\partial y^\alpha}{\partial x^r}\cdots\frac{\partial y^\delta}{\partial x^k} + \omega_{rj}\,\omega_{ik} - \omega_{rk}\,\omega_{ij},$$

where we have put

(57.15) $$B_{rijk} = g_{rh}\,B^h_{ijk}, \quad \overline{B}_{\alpha\beta\gamma\delta} = a_{\alpha\sigma}\,\overline{B}^\sigma_{\beta\gamma\delta}.$$

In the present case equations (55.7) become

(57.16) $$\omega_{ij;k} - \omega_{ik;j} = \overline{B}_{\alpha\beta\gamma\delta}\,\nu^\alpha\frac{\partial y^\beta}{\partial x^i}\cdots\frac{\partial y^\delta}{\partial x^k}.$$

If we substitute in these equations the expressions (57.4) for ω_{ij}, we have that equations (55.8) are equivalent to

$$
\begin{aligned}
g_{ir;k}\, l^r_j - g_{ir;j}\, l^r_k &= \bar{B}^{\alpha}_{\beta\gamma\delta} \frac{\partial y^\gamma}{\partial x^j} \frac{\partial y^\delta}{\partial x^k} \left(\nu_\alpha \frac{\partial y^\beta}{\partial x^i} + g_{ir}\, \nu^\beta \frac{\partial x^r}{\partial y^\alpha} \right) \\
&= \bar{B}_{\alpha\beta\gamma\delta} \frac{\partial y^\gamma}{\partial x^j} \frac{\partial y^\delta}{\partial x^k} \left(\nu^\alpha \frac{\partial y^\beta}{\partial x^i} + \nu^\beta \frac{\partial y^\alpha}{\partial x^i} \right),
\end{aligned} \tag{57.17}
$$

the last expression being a consequence of (57.6), (57.7) and (57.13). To these equations must be added (55.9) which reduce to

$$
g^{hi} (\omega_{hj}\, \omega_{ik} - \omega_{hk}\, \omega_{ij}) + \bar{B}_{\alpha\beta\gamma\delta}\, \nu^\alpha \nu^\beta \frac{\partial y^\gamma}{\partial x^j} \frac{\partial y^\delta}{\partial x^k} = 0. \tag{57.18}
$$

In a Riemannian space V_{n+1} for which $a_{\alpha\beta}$ is the fundamental tensor, it and g_{ij} are symmetric. Also $g_{ij;k} = 0$ and $\bar{B}_{\alpha\beta\gamma\delta}$ is skew-symmetric in α and β. Consequently in this case equations (57.17) and (57.18) are satisfied identically and (57.14) and (57.16) assume the form of the Gauss and Codazzi equations.*

58. Parallelism and associate directions in a hypersurface. Let λ^i be the components in the x's of a field of vectors in V_n and ξ^α the components in the y's of this vector-field in V_{n+1}. Then we have

$$
\xi^\alpha = \lambda^i \frac{\partial y^\alpha}{\partial x^i}. \tag{58.1}
$$

If we differentiate these equations with respect to x^j, we have, in consequence of (55.2),

$$
\xi^\alpha{}_{,\beta} \frac{\partial y^\beta}{\partial x^j} = \lambda^i{}_{;j} \frac{\partial y^\alpha}{\partial x^i} - \omega_{ij}\, \lambda^i \nu^\alpha. \tag{58.2}
$$

At points of a curve C in V_n, whose coördinates are expressed in terms of a parameter t, we have

$$
\xi^\alpha{}_{,\beta} \frac{d y^\beta}{d t} = \lambda^i{}_{;j} \frac{d x^j}{d t} \frac{\partial y^\alpha}{\partial x^i} - \omega_{ij}\, \lambda^i \frac{d x^j}{d t} \nu^\alpha. \tag{58.3}
$$

* 1926, 1, p. 149.

These equations may be written in the form

$$(58.4) \qquad \eta^\alpha = \mu^i \frac{\partial y^\alpha}{\partial x^i} - \omega_{ij} \lambda^i \frac{dx^j}{dt} \nu^\alpha ,$$

where η^α and μ^i are respectively associate directions (§ 16) of the vector in V_{n+1} and V_n with respect to C.

If the vectors are parallel in V_{n+1} with respect to C, $\eta^\alpha = f(t)\, \xi^\alpha$ (cf. § 7) and from (58.4) and (58.1) it follows that

$$(58.5) \qquad \mu^i = f(t)\, \lambda^i, \qquad \omega_{ij} \lambda^i \frac{dx^j}{dt} = 0.$$

Hence we have:

When a family of contravariant vectors in a hypersurface are parallel in the enveloping space with respect to a curve, they are parallel in the hypersurface with respect to the curve, for every choice of the vector ν^α.

From (58.4) and (58.5) it follows that a necessary and sufficient condition that a family of vectors in V_n at points of a curve C be parallel in V_n with respect to C, when they are not parallel in V_{n+1}, is that there exist a function $f(t)$ such that

$$(58.6) \qquad \eta^\alpha - f(t)\, \xi^\alpha = - \omega_{ij} \lambda^i \frac{dx^j}{dt} \nu^\alpha .$$

Conversely, if a vector field λ^i is such that $\omega_{ij} \lambda^i \dfrac{dx^j}{dt} = 0$ and (58.6) hold along a curve, it follows from (58.1), (58.3), (58.4) and (58.6) that

$$\left(\lambda^i_{\ ;j} \frac{dx^j}{dt} - f(t)\, \lambda^i\right) \frac{\partial y^\alpha}{\partial x^i} = 0.$$

Since the rank of the matrix $\left\| \dfrac{\partial y^\alpha}{\partial x^i} \right\|$ is n, we have that the vectors of the field are parallel in V_n with respect to the curve.

Two hypersurfaces V_n and $\overline{V}_n$ are said to be *tangent* along a curve C, if they have the same covariant pseudonormal (§ 50) at points of C. Since the components of the pseudonormal are determined only to within a factor, there is no loss of generality in assuming that ω_{ij} and ν^α are the same for V_n and $\overline{V}_n$ at points of C. Hence from the above results we have:

*When two hypersurfaces are tangent along a curve, contravariant vectors parallel in one with respect to the curve are parallel in the other.**

59. Curvature of a curve in a hypersurface. When in equations (58.3) we take for ξ^α the tangent vector to C, that is, $\xi^\alpha = \dfrac{dy^\alpha}{dt}$, these equations become

$$\begin{aligned}(59.1)\qquad &\frac{d^2 y^\alpha}{dt^2} + \overline{\Gamma}^\alpha_{\beta\gamma}\frac{dy^\beta}{dt}\frac{dy^\gamma}{dt}\\ &= \left(\frac{d^2x^i}{dt^2} + \Gamma^i_{jk}\frac{dx^j}{dt}\frac{dx^k}{dt}\right)\frac{dy^\alpha}{dx^i} - \omega_{ij}\frac{dx^i}{dt}\frac{dx^j}{dt}\nu^\alpha.\end{aligned}$$

From these equations and (7.6) we have:

When a path of a space lies in a hypersurface, it is a path of the latter, for an arbitrary choice of ν^α; and it is a curve for which

$$(59.2)\qquad \omega_{ij}\, dx^i\, dx^j = 0.$$

This is a corollary of the first theorem of § 58.

If C is not a path, we choose for the parameter an affine parameter s of the path tangent to C at a point P. Then at P we have

$$(59.3)\qquad \eta^\alpha = \overline{\eta}^\alpha - \frac{\nu^\alpha}{R},$$

where η^α are the components of the first curvature vector of C at P for V_{n+1} (§ 24).

$$(59.4)\qquad \frac{1}{R} = \omega_{ij}\frac{dx^i}{ds}\frac{dx^j}{ds}$$

and

$$(59.5)\qquad \overline{\eta}^\alpha = \mu^i\frac{\partial y^\alpha}{\partial x^i}, \qquad \frac{d^2x^i}{ds^2} + \Gamma^i_{jk}\frac{dx^j}{ds}\frac{dx^k}{ds} = \mu^i.$$

* Cf. 1926, 1, p. 75.

The vector μ^i is in the pencil determined by the tangent to C and its first curvature vector in V_n. If it is tangent to C, then C is a path in V_n and ν^α is a vector of the pencil determined by this tangent and the first curvature vector in V_{n+1}.

When ν^α is the contravariant pseudonormal to V_n (§ 56), we call $1/R$ the *normal curvature* of the curve, and $\bar{\eta}^\alpha$ the *relative curvature vector* of the curve in V_n. If $a_{\alpha\beta}\,\eta^\alpha\,\eta^\beta$ and $a_{\alpha\beta}\,\bar{\eta}^\alpha\,\bar{\eta}^\beta$ are not zero (§ 57), we put

$$(59.6) \qquad |a_{\alpha\beta}\,\eta^\alpha\,\eta^\beta| = \frac{1}{\varrho^2}, \qquad |a_{\alpha\beta}\,\bar{\eta}^\alpha\,\bar{\eta}^\beta| = \frac{1}{\varrho_r^2},$$

and call $1/\varrho$ the *first curvature* of C in V_{n+1} and $1/\varrho_r$ the *relative curvature* of C in V_n, as in the case of a Riemannian space.*

60. Asymptotic lines, conjugate directions and lines of curvature of a hypersurface. The associate covariant vector in a space V_{n+1} (§ 16) of the pseudonormal ν_α of a V_n with respect to a curve C of V_n is given by

$$\nu_{\alpha,\beta}\,\frac{dy^\beta}{dt} = \mu_\alpha.$$

From these equations and (54.1) we have:

A necessary and sufficient condition that the associate covariant vector of the covariant pseudonormal to a hypersurface with respect to a curve of the latter be pseudoörthogonal to the curve at a point is that the direction of the curve satisfy the condition

$$(60.1) \qquad \omega_{ij}\,dx^i\,dx^j = 0.$$

As this is a property of asymptotic directions of a hypersurface of a Riemannian space,† we call the directions defined by (60.1) the *asymptotic directions*. A curve whose direction at every point is asymptotic we call an *asymptotic line*. We

* Cf. 1926, 1, p. 151.

† 1926, 1, p. 157.

note that asymptotic lines and asymptotic directions are independent of the choice of the vector ν^α.*

From (59.1) and (59.3) we have the theorems:

When an asymptotic line is a path of the hypersurface, it is a path of the enveloping space and conversely.

The first curvature vector, in a space, of a curve in a hypersurface is tangential to the hypersurface at a point, when, and only when, the direction of the curve at the point is asymptotic.

If C is a path of V_n and s is an affine parameter in V_n for the path, equations (59.1) become

$$\frac{d^2 y^\alpha}{ds^2} + \overline{\Gamma}^\alpha_{\beta\gamma} \frac{dy^\beta}{ds} \frac{dy^\gamma}{ds} = -\omega_{ij} \frac{dx^i}{ds} \frac{dx^j}{ds} \nu^\alpha.$$

Hence we have:

A path of a hypersurface in an asymptotic direction at a point has contact of the second or higher order at the point with the path of the enveloping space through the point in this direction.

As in Riemannian geometry, we say that two directions at a point of a hypersurface are *conjugate*, if

$$\omega_{ij}\, dx^i\, \delta x^j = 0.$$

Thus asymptotic directions are self-conjugate. From § 58 we have:

In order that a family of vectors at points of a curve of a hypersurface be parallel both with respect to the hypersurface and the enveloping space, it is necessary that the direction of the vectors be conjugate to the curve.

If, whenever possible, the vector ν^α is chosen in such a manner that the vector l_i is zero, we have along any curve C, in consequence of (55.3)

$$\nu^\alpha{}_{,\beta} \frac{dy^\beta}{dt} = l^i_j \frac{\partial y^\alpha}{\partial x^i} \frac{dx^j}{dt}. \tag{60.2}$$

The left-hand member of this equation is the associate direction in V_{n+1} of the vector ν^α with respect to the curve. In order

* Cf. *Schouten*, 1924, 1, p. 148.

that this direction may coincide with the tangent to the curve, we must have

$$l^i_j \frac{\partial y^\alpha}{\partial x^i} \frac{dx^j}{dt} = \frac{1}{\varrho} \frac{dy^\alpha}{dt}.$$

Since the rank of the matrix $\left\| \frac{\partial y^\alpha}{\partial x^i} \right\|$ in n, these equations are equivalent to

$$(\varrho l^i_j - \delta^i_j) \frac{dx^j}{dt} = 0. \tag{60.3}$$

Conversely for each root of the determinant equation

$$|\varrho l^i_j - \delta^i_j| = 0 \tag{60.4}$$

a direction is determined for which the associate direction of ν^α in V_{n+1} coincides with this direction. When in particular the conditions of § 56 are satisfied, ν^α is the contravariant pseudonormal and the curves of V_n defined by (60.3) are an evident generalization of the lines of curvature in a Riemannian space.* Accordingly we call the curves defined by (60.3) the *lines of curvature* of V_n.

If the roots of (60.4) are real and distinct, there are n uniquely determined families of real lines of curvature. If ϱ is a real root of order r, it is possible to find r linearly independent families of lines of curvature corresponding to this root; moreover, any family of curves linearly dependent upon these families also satisfies (60.3).

Each choice of a vector ν^α determines a tensor l^i_j and consequently leads to equations of the form (60.3). However, if $l_i \neq 0$, the associate direction of the vector ν^α satisfies the above condition only in case $l_i \frac{dx^i}{dt} = 0$, as follows from (55.3). We reserve the term lines of curvature for the case when $l_i = 0$.†

If equations (60.3) are multiplied by g_{ki} (cf. § 57) and summed for i, we have from (57.4)

* 1926, 1, p. 157.

† Cf. *Schouten*, 1924, 1, p. 148.

$$(60.5)\qquad (\omega_{kj}\,\varrho - g_{kj})\frac{dx^j}{dt} = 0.$$

If as in § 59 we use for parameter an affine parameter s of the path tangent to a line of curvature at a point and put $g_{jk}\frac{dx^j}{ds}\frac{dx^k}{ds} = \varphi$, then $\varrho = R\varphi$. A discussion of equations (60.5) can be made similar to that of the corresponding equations for a hypersurface of a Riemannian space.* In particular, if λ_1^i and λ_2^i are the directions defined by (60.5) for two distinct roots of (60.4), we have

$$g_{ij}\,\lambda_1^i\,\lambda_2^j = 0 \qquad \omega_{ij}\,\lambda_1^i\,\lambda_2^j = 0.$$

From the second of these equations it follows that the two directions are conjugate.

61. Projectively flat spaces for which B_{ij} is symmetric. Consider for a space V_n with a symmetric connection the system of equations

$$(61.1)\qquad \theta_{,ij} = a_{ij}\theta.$$

The conditions of integrability (6.4) of these equations are reducible by means (61.1) to

$$\theta_{,h}(B^h_{ijk} + \delta^h_j\,a_{ik} - \delta^h_k\,a_{ij}) - \theta\,(a_{ij,k} - a_{ik,j}) = 0.$$

In order that equations (61.1) be completely integrable, and consequently that the general solution admit $n+1$ arbitrary constants, it is necessary that

$$B^h_{ijk} + \delta^h_j\,a_{ik} - \delta^h_k\,a_{ij} = 0, \qquad a_{ij,k} - a_{ik,j} = 0.$$

Contracting the first for h and k, we have

$$(61.2)\qquad a_{ij} = \frac{1}{n-1}B_{ij},$$

so that the above conditions are

$$(61.3)\quad B^h_{ijk} + \frac{1}{n-1}(\delta^h_j\,B_{ik} - \delta^h_k\,B_{ij}) = 0, \quad B_{ij,k} - B_{ik,j} = 0.$$

* Cf. 1926, 1, p. 153.

Hence V_n is a projectively flat space (§ 34) and B_{ij} is symmetric as follows from (61.1), (61.2) and (6.3).

Since each solution of (61.1) is determined by initial values of θ and $\frac{\partial \theta}{\partial x^i}$, there exist $n+1$ solutions $\theta^\alpha(x^1, \cdots, x^n)$ for which the determinant

$$\text{(61.4)} \qquad \Delta = \begin{vmatrix} \frac{\partial \theta^1}{\partial x^1} & \cdots & \frac{\partial \theta^1}{\partial x^n} & \theta^1 \\ \cdot & \cdot & \cdot & \cdot \\ \cdot & \cdot & \cdot & \cdot \\ \frac{\partial \theta^{n+1}}{\partial x^1} & \cdots & \frac{\partial \theta^{n+1}}{\partial x^n} & \theta^{n+1} \end{vmatrix}$$

is not identically zero and the matrix of the first n columns is of rank n. Hence the jacobian of the equations

$$\text{(61.5)} \qquad y^\alpha = e^{x^{n+1}} \theta^\alpha(x^1, \cdots, x^n) \qquad (\alpha = 1, \cdots, n+1)$$

is not zero, and these equations define a transformation of coördinates in a space V_{n+1}. We define a connection for this V_{n+1} in the coördinates x^α, by taking for $\Gamma^i_{jk} (i, j, k = 1, \cdots, n)$ the expressions for these functions for the given V_n, and in addition

$$\text{(61.6)} \quad \Gamma^{n+1}_{ij} = \frac{1}{n-1} B_{ij}, \quad \Gamma^\alpha_{n+1\,\beta} = \delta^\alpha_\beta, \quad (\alpha, \beta = 1, \cdots, n+1).$$

If $\overline{\Gamma}^\alpha_{\beta\gamma}$ denote the coefficients of the connection in the y's, it follows from the equations

$$\frac{\partial^2 y^\alpha}{\partial x^\beta \partial x^\gamma} + \overline{\Gamma}^\alpha_{\mu\nu} \frac{\partial y^\mu}{\partial x^\beta} \frac{\partial y^\nu}{\partial x^\gamma} = \Gamma^\mu_{\beta\gamma} \frac{\partial y^\alpha}{\partial x^\mu}$$
$$(\alpha, \beta, \gamma, \mu, \nu = 1, \cdots, n+1)$$

and from (61.1), (61.2) and (61.5) that

$$\overline{\Gamma}^\alpha_{\mu\nu} \frac{\partial y^\mu}{\partial x^\beta} \frac{\partial y^\nu}{\partial x^\gamma} = 0.$$

Consequently $\overline{\Gamma}^\alpha_{\mu\nu} = 0$ and V_{n+1} is a euclidean, or flat space. From the definition of the affine connection in V_{n+1} it follows that the induced connection in the hypersurface $x^{n+1} = 0$ is that of the given space V_n. Hence we have:

*A projectively flat space of n dimensions for which B_{ij} is symmetric can be immersed in a flat space of $n+1$ dimensions.**

In order to investigate the situation more fully, we observe that by suitable linear combinations, with real coefficients, of the θ's the fundamental form of V_{n+1} is reducible to $\sum_\alpha e_\alpha (dy^\alpha)^2$, where the e's are plus or minus one. If we denote by $a_{\alpha\beta}$ the coefficients of this form in the x's, we have

$$(61.7)\qquad \begin{gathered} a_{ij} = e^{2x^{n+1}} \sum_\alpha e_\alpha \frac{\partial \theta^\alpha}{\partial x^i} \frac{\partial \theta^\alpha}{\partial x^j}, \quad a_{i\,n+1} = e^{2x^{n+1}} \sum_\alpha e_\alpha \theta^\alpha \frac{\partial \theta^\alpha}{\partial x^i}; \\ a_{n+1\,n+1} = e^{2x^{n+1}} \sum_\alpha e_\alpha (\theta^\alpha)^2. \end{gathered}$$

If we put

$$(61.8)\qquad \sum_\alpha e_\alpha (\theta^\alpha)^2 = 2\lambda,$$

the successive covariant derivatives of this equation are reducible by means of (61.1) and (61.2) to

$$(61.9)\qquad \begin{gathered} \sum_\alpha e_\alpha \theta^\alpha \theta^\alpha_{,i} = \lambda_{,i}, \\ \frac{2\lambda}{n-1} B_{ij} + \sum_\alpha e_\alpha \theta^\alpha_{,i} \theta^\alpha_{,j} = \lambda_{,ij}. \end{gathered}$$

From the results of § 53 it follows that the coefficients of the induced connection in the hypersurface $x^{n+1} = 0$ are the Christoffel symbols of the second kind formed with respect to

$$g_{ij} = \sum_\alpha e_\alpha \theta^\alpha_{,i} \theta^\alpha_{,j}$$

only in case $a_{i\,n+1} = 0$, that is, when λ is a constant. In this case we have

$$B_{ij} = \frac{1-n}{2\lambda} g_{ij},$$

and consequently the hypersurface is of constant Riemannian curvature.†

* *Eisenhart*, 1926, 9, p. 338.

† 1926, 1, pp. 135, 203.

Consider now any hypersurface of a flat space V_{n+1} for which the coördinates y^α are cartesian and the fundamental form is $\sum_\alpha e_\alpha (dy^\alpha)^2$, the equation of the hypersurface being

$$F(y^1, \cdots, y^{n+1}) = 0. \tag{61.10}$$

This equation may be replaced by

$$y^\alpha = \theta^\alpha (x^1, \cdots, x^n) \tag{61.11}$$

where the functions θ^α are arbitrary, except that (61.10) is satisfied and the jacobian of any n of them is not zero. When, and only when, the function F is not homogeneous in the y's, the determinant Δ defined by (61.4) is different from zero; that is, when the hypersurface is not a hypercone with vertex at the origin, or a hyperplane through the origin. Consequently, with these exceptions, the functions θ^α can be chosen so that equations (61.5) define a transformation of coördinates in V_{n+1} such that $x^{n+1} = 0$ is the equation of the given hypersurface. The coefficients of the induced connection in the given hypersurface are given by

$$\Gamma^i_{jk} = \begin{Bmatrix} i \\ jk \end{Bmatrix}_a, \tag{61.12}$$

where the latter are defined with respect to the a's given by (61.7). Since the Christoffel symbols formed with respect to the y's are zero, we have

$$\begin{Bmatrix} \alpha \\ \beta\gamma \end{Bmatrix}_a = \frac{\partial^2 y^\delta}{\partial x^\beta \partial x^\gamma} \frac{\partial x^\alpha}{\partial y^\delta}. \tag{61.13}$$

From these equations and (61.12) we have

$$\frac{\partial^2 y^\alpha}{\partial x^i \partial x^j} - \Gamma^h_{ij} \frac{\partial y^\alpha}{\partial x^h} = \begin{Bmatrix} n+1 \\ ij \end{Bmatrix}_a y^\alpha, \tag{61.14}$$

which reduce to the form (61.1) because of (61.5). Consequently

$$B_{ij} = (n-1) \begin{Bmatrix} n+1 \\ ij \end{Bmatrix}_a,$$

and for the induced connection the hypersurface is projectively flat. Also from (61.5) it follows that

$$\frac{\partial y^\alpha}{\partial x^{n+1}} = y^\alpha,$$

by means of which and (61.14) we get the equations (61.6).

Returning to the consideration of the cases excepted above, we see that by a suitable change of origin of the y's, the hypersurfaces excepted for one coördinate system are not excepted for another. Hence we have:

The vector-field ν^α can be chosen at points of any hypersurface of a flat space so that for the induced connection the hypersurface is projectively flat.

62. Covariant pseudonormals to a sub-space. When a space V_m is referred to coördinates y^α, a sub-space, or sub-variety, V_n is defined by

$$\varphi^\sigma(y^1, \cdots, y^m) = 0 \qquad (\sigma = n+1, \cdots, m). \tag{62.1}$$

If we put

$$x^i = \varphi^i(y^1, \cdots, y^m), \qquad x^\sigma = \varphi^\sigma, * \tag{62.2}$$

where the functions φ^i are arbitrary except that the jacobian of the m φ's is different from zero, equations (62.2) define a coördinate system for which the given V_n is defined by the equations $x^\sigma = 0$ $(\sigma = n+1, \cdots, m)$.

Any displacement in V_n satisfies the conditions

$$\frac{\partial \varphi^\sigma}{\partial y^\alpha} dy^\alpha = 0, \tag{62.3}$$

and consequently the covariant vectors $\nu_\alpha^{(\sigma)}$ in V_m defined by

$$\nu_\alpha^{(\sigma)} = \frac{\partial \varphi^\sigma}{\partial y^\alpha} = \frac{\partial x^\sigma}{\partial y^\alpha} \tag{62.4}$$

* In this and the following sections latin indices take the values $1, \cdots, n$, letters at the beginning of the greek alphabet values $1, \cdots, m$ and those at the end $n+1, \cdots, m$.

are pseudoörthogonal at any point of V_n to any displacement in the latter. We call them *covariant pseudonormals* to V_n; evidently any linear combination of them is the most general covariant pseudonormal to V_n.

If we put

$$\nu^\alpha_{(\sigma)} = \frac{\partial y^\alpha}{\partial x^\sigma}, \tag{62.5}$$

$\nu^\alpha_{(\sigma)}$, for a given value of σ, at any point are the components of the contravariant vector tangential to the curve of parameter x^σ at the point, that is, the curve along which all the x's but x^σ are constant. As thus defined the functions $\nu^\alpha_{(\sigma)}$ depend upon the choice of the functions φ^i in (62.2), whereas $\nu^{(\sigma)}_\alpha$ do not. From (62.4) and (62.5) we have

$$\nu^\alpha_{(\sigma)}\, \nu^{(\tau)}_\alpha = \delta^\tau_\sigma \tag{62.6}$$

and

$$\nu^\alpha_{(\sigma)} \frac{\partial x^i}{\partial y^\alpha} = 0. \tag{62.7}$$

When equations (62.2) are solved for the y's, we have

$$y^\alpha = f^\alpha(x^1, \cdots, x^m) \tag{62.8}$$

or as power series in $x^{n+1}, \cdots, x^m$

$$y^\alpha = f^\alpha_0(x^1, \cdots, x^n) + f^\alpha_\sigma(x^1, \cdots, x^n)\, x^\sigma + \cdots. \tag{62.9}$$

Consequently V_n is defined by the parametric equations

$$y^\alpha = f^\alpha_0(x^1, \cdots, x^n) \tag{62.10}$$

and at points of V_n

$$\nu^\alpha_{(\sigma)} = f^\alpha_\sigma(x^1, \cdots, x^n). \tag{62.11}$$

63. Derived tensors in a sub-space. Induced affine connection. For a tensor in V_m the quantities given by (52.3), (52.11) and (52.15), where greek indices take the values $1, \cdots, m$, evaluated at points of V_n define a *derived* tensor in V_n. In particular, the derived vector of any covariant pseudonormal is zero, and the derived vector of any $\nu^{\alpha}_{(\sigma)}$ is zero.

If we put

$$B^{\alpha}_{\beta} = \delta^{\alpha}_{\beta} - \nu^{\alpha}_{(\sigma)} \nu^{(\sigma)}_{\beta}, \tag{63.1}$$

we have

$$B^{\alpha}_{\beta} \frac{\partial x^i}{\partial y^{\alpha}} = \frac{\partial x^i}{\partial y^{\beta}}, \qquad B^{\alpha}_{\beta} \frac{\partial x^{\tau}}{\partial y^{\alpha}} = 0,$$

$$B^{\alpha}_{\beta} \frac{\partial y^{\beta}}{\partial x^i} = \frac{\partial y^{\alpha}}{\partial x^i}, \qquad B^{\alpha}_{\beta} \frac{\partial y^{\beta}}{\partial x^{\tau}} = 0.$$

Hence with this interpretation of B^{α}_{β} equations (52.12), (52.14) and (52.16) define tensors in V_m, *associate* to the given tensors, such that in the x's any component involving one or more indices $n+1, \cdots, m$ is zero and the other components are those of the derived tensor. Furthermore, the last theorem of § 52 holds for a V_n in a V_m.

If $\Gamma^{\alpha}_{\beta\gamma}$ and $\bar{\Gamma}^{\alpha}_{\beta\gamma}$ are the coefficients of an affine connection in V_m in the x's and y's respectively, we have equations of the form (53.1). If $\Gamma'^{\alpha}_{\beta\gamma}$ are the coefficients in coördinates x'^{α}, where

$$x'^i = \psi^i(x^1, \cdots, x^n), \qquad x'^{\sigma} = x^{\sigma},$$

then equations (53.2) hold, and thus the quantities Γ^i_{jk} and Γ'^i_{jk} evaluated at points of V_n are the coefficients of the same connection, which we say is *induced* in V_n. From the form of equations (53.1) it is seen that this induced connection varies with the choice of the vectors $\nu^{\alpha}_{(\sigma)}$. In what follows we indicate by a semi-colon followed by indices covariant differentation with respect to the induced connection. We remark that for a V_n in a V_m there is a theorem similar to the last of § 53.

64. Fundamental derived tensors in a sub-space. We denote by $\omega^{(\sigma)}_{ij}$ the tensor in V_n derived from the tensor $\nu^{(\sigma)}_{\alpha,\beta}$ in V_m, that is,

$$\omega^{(\sigma)}_{ij} = \nu^{(\sigma)}_{\alpha,\beta} \frac{\partial y^\alpha}{\partial x^i} \frac{\partial y^\beta}{\partial x^j}. \tag{64.1}$$

In the x's the components of the vectors $\nu^{(\sigma)}_\alpha$ and $\nu^\alpha_{(\sigma)}$ are

$$\nu^{(\sigma)}_\alpha = \delta^\sigma_\alpha, \qquad \nu^\alpha_{(\sigma)} = \delta^\alpha_\sigma, \tag{64.2}$$

as follow from (62.4) and (62.5). Consequently in the x's equations (64.1) are

$$\omega^{(\sigma)}_{ij} = - \Gamma^\sigma_{ij}. \tag{64.3}$$

We denote by $l^i_{(\sigma)j}$ $l^{(\tau)}_{(\sigma)i}$ the tensors in V_n derived from $\nu^\alpha_{(\sigma),\beta}$ and $\nu^\alpha_{(\sigma),\beta}\,\nu^{(\tau)}_\alpha$ in V_m, that is,

$$l^i_{(\sigma)j} = \nu^\alpha_{(\sigma),\beta} \frac{\partial x^i}{\partial y^\alpha} \frac{\partial y^\beta}{\partial x^j} \tag{64.4}$$

and

$$l^{(\tau)}_{(\sigma)i} = \nu^\alpha_{(\sigma),\beta}\, \nu^{(\tau)}_\alpha \frac{\partial y^\beta}{\partial x^i}. \tag{64.5}$$

In the x's the components of these tensors are

$$l^i_{(\sigma)j} = \Gamma^i_{\sigma j}, \qquad l^{(\tau)}_{(\sigma)i} = \Gamma^\tau_{\sigma i}. \tag{64.6}$$

In consequence of (62.6) we have from (64.5)

$$l^{(\tau)}_{(\sigma)i} = - \nu^{(\tau)}_{\alpha,\beta}\, \nu^\alpha_{(\sigma)} \frac{\partial y^\beta}{\partial x^i}. \tag{64.7}$$

The geometrical significance of these tensors is pointed out in the next section.

In order to study the effect of a change of the vectors $\nu^\alpha_{(\sigma)}$, we consider a transformation of coördinates of the form

$$x'^i = x^i + \psi^i_\sigma x^\sigma, \qquad x'^\sigma = x^\sigma, \tag{64.8}$$

where the ψ's are functions of $x^1, \cdots, x^n$. At points of V_n we have

$$\frac{\partial y^\alpha}{\partial x^i} = \frac{\partial y^\alpha}{\partial x'^\beta}\frac{\partial x'^\beta}{\partial x^i} = \frac{\partial y^\alpha}{\partial x'^i},$$

(64.9)
$$\frac{\partial y^\alpha}{\partial x^\sigma} = \frac{\partial y^\alpha}{\partial x'^\sigma} + \frac{\partial y^\alpha}{\partial x'^i}\psi^i_\sigma, \qquad \frac{\partial x'^i}{\partial y^\alpha} = \frac{\partial x^i}{\partial y^\alpha} + \psi^i_\sigma \frac{\partial x}{\partial y^\alpha},$$

$$\frac{\partial x'^\sigma}{\partial y^\alpha} = \frac{\partial x^\sigma}{\partial y^\alpha}, \qquad \frac{\partial^2 y^\alpha}{\partial x^i \partial x^j} = \frac{\partial^2 y^\alpha}{\partial x'^i \partial x'^j},$$

$$\frac{\partial^2 y^\alpha}{\partial x^i \partial x^\sigma} = \frac{\partial^2 y^\alpha}{\partial x'^i \partial x'^j}\psi^j_\sigma + \frac{\partial^2 y^\alpha}{\partial x'^i \partial x'^\sigma} + \frac{\partial y^\alpha}{\partial x'^j}\frac{\partial \psi^j_\sigma}{\partial x^i}.$$

From the fourth of these equations we have that $\nu'^{(\sigma)}_\alpha = \nu^{(\sigma)}_\alpha$, that is, the covariant pseudonormals are unaltered. The second and last sets of these equations are reducible by means of the others to

(64.10)
$$\nu^\alpha_{(\sigma)} = \nu'^\alpha_{(\sigma)} + \frac{\partial y^\alpha}{\partial x'^i}\psi^i_\sigma$$

and

$$\frac{\partial \nu'^\alpha_{(\sigma)}}{\partial x'^i} = \frac{\partial \nu^\alpha_{(\sigma)}}{\partial x^i} - \frac{\partial^2 y^\alpha}{\partial x^i \partial x^j}\psi^j_\sigma - \frac{\partial y^\alpha}{\partial x^j}\frac{\partial \psi^j_\sigma}{\partial x^i}.$$

From these two sets of equations, (64.9) and (53.1) we have

(64.11)
$$\begin{aligned}\nu'^\alpha_{(\sigma),\beta}\frac{\partial y^\beta}{\partial x'^i} &= \nu^\alpha_{(\sigma),\beta}\frac{\partial y^\beta}{\partial x^i} - \left(\frac{\partial^2 y^\alpha}{\partial x^i \partial x^j} + \Gamma^\alpha_{\beta\gamma}\frac{\partial y^\beta}{\partial x^i}\frac{\partial y^\gamma}{\partial x^j}\right)\psi^j_\sigma \\ &\quad - \frac{\partial y^\alpha}{\partial x^j}\frac{\partial \psi^j_\sigma}{\partial x^i} \\ &= \nu^\alpha_{(\sigma),\beta}\frac{\partial y^\beta}{\partial x^i} - \psi^j_{\sigma;i}\frac{\partial y^\alpha}{\partial x^j} + \psi^j_\sigma\,\omega^{(\tau)}_{ij}\,\nu^\alpha_{(\tau)}.\end{aligned}$$

If we define $l'^i_{(\sigma)j}$ and $l'^{(\tau)}_{(\sigma)i}$ by equations analogous to (64.4) and (64.5), we have

(64.12)
$$l'^i_{(\sigma)j} = l^i_{(\sigma)j} - \psi^i_{\sigma;j} + \psi^i_\varrho\, l^{(\varrho)}_{(\sigma)j} + \psi^k_\sigma\,\psi^i_\varrho\,\omega^{(\varrho)}_{jk}$$

(64.13)
$$l'^{(\tau)}_{(\sigma)j} = l^{(\tau)}_{(\sigma)j} + \psi^k_\sigma\,\omega^{(\tau)}_{jk}.$$

There is also another element of indeterminateness due to the fact that the covariant pseudonormals are not uniquely determined. In fact, the sub-space defined by (62.1) is likewise defined by the equations $\bar{\varphi}^\sigma = 0$, where $\bar{\varphi}^\sigma = A^\sigma_\tau \varphi^\tau$, the A's being arbitrary functions of the y's subject to the condition that the determinant $|A^\sigma_\tau|$ is not zero as a consequence of (62.1). From the above we have

$$\bar{\nu}^{(\sigma)}_{\alpha,\beta} = \bar{\varphi}^\sigma_{\ ,\alpha\beta} = A^\sigma_\tau \nu^{(\tau)}_{\alpha,\beta} + A^\sigma_{\tau,\beta} \nu^{(\tau)}_\alpha + A^\sigma_{\tau,\alpha} \nu^{(\tau)}_\beta + A^\sigma_{\tau,\alpha\beta} \varphi^\tau.$$

From these equations and (64.1) we have at points of V_n

$$\bar{\omega}^{(\sigma)}_{ij} = A^\sigma_\tau \omega^{(\tau)}_{ij}, \tag{64.14}$$

where now A^σ_τ may be interpreted as arbitrary functions of $x^1, \cdots, x^n$ such that the determinant $|A^\sigma_\tau|$ is not zero. An application of the foregoing results will be made in the next section.

65. Generalized equations of Gauss and Codazzi. At points of a sub-space V_n of a V_m as defined in § 62, equations (55.1) may be written by means of (64.3)

$$y^\alpha_{\ ;ij} + \bar{\Gamma}^\alpha_{\beta\gamma} \frac{\partial y^\beta}{\partial x^i} \frac{\partial y^\gamma}{\partial x^j} = -\omega^{(\sigma)}_{ij} \nu^\alpha_{(\sigma)}. \tag{65.1}$$

When in equations (55.1) we replace j by σ, the resulting equations may be written

$$\frac{\partial \nu^\alpha_{(\sigma)}}{\partial x^i} + \bar{\Gamma}^\alpha_{\beta\gamma} \nu^\beta_{(\sigma)} \frac{\partial y^\gamma}{\partial x^i} = \Gamma^\beta_{\sigma i} \frac{\partial y^\alpha}{\partial x^\beta},$$

and at points of V_n these equations become, in consequence of (64.6),

$$\frac{\partial \nu^\alpha_{(\sigma)}}{\partial x^i} + \bar{\Gamma}^\alpha_{\beta\gamma} \nu^\beta_{(\sigma)} \frac{\partial y^\gamma}{\partial x^i} = l^j_{(\sigma)i} \frac{\partial y^\alpha}{\partial x^j} + l^{(\tau)}_{(\sigma)i} \nu^\alpha_{(\tau)}. \tag{65.2}$$

With the aid of the identities (55.4) for the present case, we obtain as the conditions of integrability of equations (65.1)

$$\frac{\partial y^\alpha}{\partial x^h}(B^h_{ijk} + \overset{(\sigma)}{\omega}_{ij}\, l^h_{(\sigma)k} - \overset{(\sigma)}{\omega}_{ik}\, l^h_{(\sigma)j}) - \overline{B}^\alpha_{\beta\gamma\delta}\frac{\partial y^\beta}{\partial x^i}\frac{\partial y^\gamma}{\partial x^j}\frac{\partial y^\delta}{\partial x^k}$$
$$+ \nu^\alpha_{(\sigma)}(\overset{(\sigma)}{\omega}_{ij;k} - \overset{(\sigma)}{\omega}_{ik;j} + \overset{(\tau)}{\omega}_{ij}\, \overset{(\sigma)}{l}_{(\tau)k} - \overset{(\tau)}{\omega}_{ik}\, \overset{(\sigma)}{l}_{(\tau)j}) = 0.$$

If these equations are multiplied by $\dfrac{\partial x^h}{\partial y^\alpha}$ and by $\nu^{(\sigma)}_\alpha$ and α is summed, we have the respective equations

$$B^h_{ijk} = \overline{B}^\alpha_{\beta\gamma\delta}\frac{\partial y^\beta}{\partial x^i}\cdots\frac{\partial x^h}{\partial y^\alpha} - \overset{(\sigma)}{\omega}_{ij}\, l^h_{(\sigma)k} + \overset{(\sigma)}{\omega}_{ik}\, l^h_{(\sigma)j}, \tag{65.3}$$

$$\overset{(\sigma)}{\omega}_{ij;k} - \overset{(\sigma)}{\omega}_{ik;j} = \overline{B}^\alpha_{\beta\gamma\delta}\frac{\partial y^\beta}{\partial x^i}\cdots\nu^{(\sigma)}_\alpha - \overset{(\tau)}{\omega}_{ij}\, \overset{(\sigma)}{l}_{(\tau)k} + \overset{(\tau)}{\omega}_{ik}\, \overset{(\sigma)}{l}_{(\tau)j}. \tag{65.4}$$

In like manner we find that the conditions of integrability of (65.2) are

$$\begin{aligned} &l^h_{(\sigma)i;j} - l^h_{(\sigma)j;i} + \overset{(\tau)}{l}_{(\sigma)i}\, l^h_{(\tau)j} - \overset{(\tau)}{l}_{(\sigma)j}\, l^h_{(\tau)i} \\ &\qquad + \overline{B}^\alpha_{\beta\gamma\delta}\,\nu^\beta_{(\sigma)}\frac{\partial y^\gamma}{\partial x^i}\frac{\partial y^\delta}{\partial x^j}\frac{\partial x^h}{\partial y^\alpha} = 0, \end{aligned} \tag{65.5}$$

$$\begin{aligned} &\overset{(\tau)}{l}_{(\sigma)i;j} - \overset{(\tau)}{l}_{(\sigma)j;i} + l^h_{(\sigma)j}\, \overset{(\tau)}{\omega}_{ih} - l^h_{(\sigma)i}\, \overset{(\tau)}{\omega}_{jh} \\ &\qquad + \overline{B}^\alpha_{\beta\gamma\delta}\,\nu^{(\tau)}_\alpha\, \nu^\beta_{(\sigma)}\frac{\partial y^\gamma}{\partial x^i}\frac{\partial y^\delta}{\partial x^j} = 0. \end{aligned} \tag{65.6}$$

These equations are evident generalizations of equations obtained by Voss and Ricci for a sub-space of a general Riemannian space*.

From (65.3) we have on contracting for h and i

$$S_{jk} = (\overline{S}_{\gamma\delta} - \overline{B}^\alpha_{\beta\gamma\delta}\,\nu^\beta_{(\sigma)}\,\nu^{(\sigma)}_\alpha)\frac{\partial y^\gamma}{\partial x^j}\frac{\partial y^\delta}{\partial x^k} - \overset{(\sigma)}{\omega}_{hj}\, l^h_{(\sigma)k} + \overset{(\sigma)}{\omega}_{hk}\, l^h_{(\sigma)j},$$

which is reducible by means of (65.6) to

$$S_{jk} = \overline{S}_{\gamma\delta}\frac{\partial y^\gamma}{\partial x^j}\frac{\partial y^\delta}{\partial x^k} + \overset{(\sigma)}{l}_{(\sigma)j;k} - \overset{(\sigma)}{l}_{(\sigma)k;j}. \tag{65.7}$$

From (64.13) it is seen that if the determinant of any one of the sets of functions $\overset{(\sigma)}{\omega}_{ij}$ is different from zero, or can be

* 1926, 1, p. 163.

made such by (64.14), the functions ψ^i_{σ}, and consequently the vectors $\nu^\alpha_{(\sigma)}$, can be chosen so that the sums $l'^{(\sigma)}_{(\sigma)i}$ are zero. Consequently we have from (65.7):

In general the vectors $\nu^\alpha_{(\sigma)}$ *can be chosen so that when the tensor* $\overline{B}_{\alpha\beta}$ *in the enveloping space is symmetric, the tensor* B_{ij} *of the induced connection in the sub-space also is symmetric.**

It is an algebraic problem to determine by means of equations (64.14) whether $m-n$ independent covariant pseudonormals can be chosen so that all of the determinants $|\omega^{(\sigma)}_{ij}|$ are different from zero. When this condition is satisfied, $m-n$ independent vectors $\nu^\alpha_{(\sigma)}$ can be determined by (64.13), so that $l'^{(\sigma)}_{(\sigma)j} = 0$, where σ is not summed. In this case, as follows from (65.2), the associate direction in V_m of each vector $\nu^\alpha_{(\sigma)}$ for a displacement in V_n does not have a component in the direction $\nu^\alpha_{(\sigma)}$ (cf. § 56). As this is a property of the normals to a sub-space of a Riemannian space,† we say that the corresponding vectors $\nu^\alpha_{(\sigma)}$ are *contravariant pseudonormals* to the sub-space. The process of determining these pseudonormals is not unique since each choice of covariant pseudonormals satisfying the desired conditions yields a set of contravariant pseudonormals.

When the connection is asymmetric (cf. § 55), the equations analogous to (65.3), (65.4), (66.5) and (65.6) are obtained from the latter on replacing B^h_{ijk} and $\overline{B}^\alpha_{\beta\gamma\delta}$ by L^h_{ijk} and $\overline{L}^\alpha_{\beta\gamma\delta}$, subtracting the term $2\,\Omega^h_{jk}\,\omega^{(\sigma)}_{ih}$ from the right-hand member of (65.4), and adding the respective terms $2\,\Omega^k_{ij}\,l^h_{(\sigma)k}$ and $2\,\Omega^k_{ij}\,l^{(\tau)}_{(\sigma)k}$ to the left-hand members of (65.5) and (65.6).

66. Parallelism in a sub-space. Curvature of a curve in a sub-space. The results of §§ 58, 59 can be generalized to the case of a general sub-space. In place of (58.3) we have

$$(66.1)\qquad \xi^\alpha{}_{,\beta}\frac{dy^\beta}{dt} = \lambda^i{}_{,j}\frac{dx^j}{dt}\frac{\partial y^\alpha}{\partial x^i} - \omega^{(\sigma)}_{ij}\,\lambda^i\frac{dx^j}{dt}\,\nu^\alpha_{(\sigma)}.$$

* Cf. *Schouten*, 1924, 1, p. 162.

† 1926, 1, p. 161, equations (47.9).

Consequently we have

When a family of contravariant vectors of a sub-space are parallel in the enveloping space with respect to a curve, they are parallel in the sub-space, for every choice of the vectors $\nu^\alpha_{(\sigma)}$.

In particular we have

When a path of a space lies in a sub-space, it is a path of the sub-space, for every choice of the vectors $\nu^\alpha_{(\sigma)}$; *it is a curve for which*

$$\omega^{(\sigma)}_{ij}\, dx^i\, dx^j = 0 \qquad (\sigma = n+1, \cdots, m).$$

From this theorem we have also:

A necessary and sufficient condition that every path of a sub-space be a path of the enveloping space is that

$$\omega^{(\sigma)}_{ij} = 0.$$

These sub-spaces are the analogues of totally-geodesic sub-spaces of a Riemannian space.*

When in (66.1) we replace ξ^α by $\frac{dy^\alpha}{dt}$, we have the equation obtained from (59.1) on replacing the last term by $\omega^{(\sigma)}_{ij}\,\nu^\alpha_{(\sigma)}\frac{dx^i}{dt}\frac{dx^j}{dt}$. Consequently $\omega^{(\sigma)}_{ij}\frac{dx^i}{dt}\frac{dx^j}{dt}$ is the component of the curvature of the curve in the direction $\nu^\alpha_{(\sigma)}$.

If we multiply (65.2) by $\frac{dx^i}{dt}$ for a curve, we have

$$\nu^\alpha_{(\sigma)|\beta} = l^j_{(\sigma)i}\frac{dx^i}{dt}\frac{\partial y^\alpha}{\partial x^j} + l^{(\tau)}_{(\sigma)i}\,\nu^\alpha_{(\tau)}\frac{dx^i}{dt}.$$

Hence $l^j_{(\sigma)i}\frac{dx^i}{dt}$ are the tangential components of the associate direction of the vector $\nu^\alpha_{(\sigma)}$ for the curve and $l^{(\tau)}_{(\sigma)i}\frac{dx^i}{dt}$ are the components in the directions $\nu^\alpha_{(\tau)}$.†

67. Projective change of induced connection. In order to determine the effect upon the induced connection of

* 1926, 1, p. 184.

† Cf. *Weyl*, 1922, 6, p. 156.

a projective change of connection in the enveloping space, replace $\bar{\Gamma}^{\alpha}_{\beta\gamma}$ in (53.1) by expressions of the form (32.1) and understand that α, β, γ take the values $1, \cdots, m$. From the resulting equations and (53.1) we have

$$\Gamma'^{i}_{jk} = \Gamma^{i}_{jk} + \delta^{i}_{j}\,\varphi_k + \delta^{i}_{k}\,\varphi_j, \tag{67.1}$$

where

$$\varphi_j = \psi_\beta \frac{\partial y^\beta}{\partial x^j}, \tag{67.2}$$

ψ_β being the vector in V_m determining the projective change. Hence we have:

When the connection of a space undergoes a projective change, the same is true of the induced connection of a sub-space, and the vector determining the latter is the derived vector of that determining the former.

From (67.2) it is evident that $m-n$ independent vectors ψ_α exist such that there is no change in the induced connection. If we take $\psi_\alpha = \nu^{\beta}_{(\sigma),\alpha}\,\nu^{(\sigma)}_{\beta}$, where σ is not summed, then $\varphi_j = l^{(\sigma)}_{(\sigma)j}$ (cf. § 65). Consequently when the vectors $\nu^{\alpha}_{(\sigma)}$ can be chosen so that $l^{(\sigma)}_{(\sigma)j} = 0$ (σ not summed), there is no projective change in the induced connection.

BIBLIOGRAPHY

This bibliography contains only the books and memoirs which are referred to in the text.

1869. 1. *Christoffel, E. B.*: Über die Transformation der homogenen Differentialausdrücke zweiten Grades. Journal für die reine und angew. Mathematik (Crelle), vol. 70, pp. 46–70.

1891. 1. *Goursat, E.*: Leçons sur l'intégration des équations aux dérivées partielles du premier ordre. Hermann, Paris.

1893. 1. *Lie, S.*: Vorlesungen über kontinuierliche Gruppen. Teubner, Leipzig.

1901. 1. *Ricci, G.* and *Levi-Civita, T.*: Méthodes de calcul différentiel absolu et leurs applications. Math. Annalen, vol. 54, pp. 125–201, 608.

1905. 1. *Fine, H. B.*: A college algebra. Ginn and Company, Boston.

1908. 1. *Wright, J. E.*: Invariants of quadratic differential forms. Cambridge Tract, No. 9.

1909. 1. *Eisenhart, L. P.*: A treatise on the differential geometry of curves and surfaces. Ginn and Company, Boston.

2. *Kowalewski, G.*: Einführung in die Determinantentheorie. Veit und Comp., Leipzig.

1917. 1. *Levi-Civita, T.*: Nozione di parallelismo in una varietà qualunque e consequente specificazione geometrica della curvatura Riemanniana. Rendiconti di Palermo, vol. 42, pp. 173–205.

1918. 1. *Bianchi, L.*: Lezioni sulla teoria dei gruppi continui finiti di trasformazioni. Spoerri, Pisa.

2. *Finsler, P.*: Über Kurven und Flächen in allgemeinen Räumen. Dissertation. Göttingen.

1921. 1. *Weyl, H.*: Space, time, matter. Translated by H. L. Brose. Methuen, London.

2. *Weyl, H.*: Zur Infinitesimalgeometrie; Einordnung der projektiven und der konformen Auffassung. Göttinger Nachrichten, 1921, pp. 99–112.

1922. 1. *Eisenhart, L. P.*: Fields of parallel vectors in the geometry of paths. Proceedings of the Nat. Acad. of Sciences, vol. 8, pp. 207–212.

2. *Eisenhart, L. P.*: Spaces with corresponding paths. Proceedings of the Nat. Acad. of Sciences, vol. 8, pp. 233–238.

3. *Veblen, O.*: Projective and affine geometry of paths. Proceedings of Nat. Acad. of Sciences, vol. 8, pp. 347–350.

4. *Eisenhart, L. P.* and *Veblen, O.*: The Riemann geometry and its generalization. Proceedings of Nat. Acad. of Sciences, vol. 8, pp. 19–23.

5. *Veblen, O.*: Normal coördinates for the geometry of paths. Proceedings of Nat. Acad. of Sciences, vol. 8, pp. 192–197.

6. *Weyl, H.*: Zur Infinitesimalgeometrie; p-dimensionale Fläche im n-dimensionalen Raum. Math. Zeitschrift, vol. 12, pp. 154–160.

7. *Fermi, E.*: Sopra i fenomeni che avvengono in vicinanza di una linea oraria. Rendiconti dei Lincei, vol. 31^1, pp. 21–23, 51–52.

1923. 1. *Veblen, O.* and *Thomas, T. Y.*: The geometry of paths. Transactions of the Amer. Math. Soc., vol. 25, pp. 551–608.

2. *Weitzenböck, R.*: Invariantentheorie. Noordhoff, Groningen.

3. *Eisenhart, L. P.*: Symmetric tensors of the second order whose first covariant derivatives are zero. Transactions of the Amer. Math. Soc., vol. 25, pp. 297–306.

4. *Eisenhart, L. P.*: The geometry of paths and general relativity. Annals of Mathematics, ser. 2, vol. 24, pp. 367–392.

5. *Cartan, E.*: Sur les variétés à connexion affine et la théorie de la relativité généralisée. Annales de L'École Norm. Super., ser. 3, vol. 40, pp. 325–412.

6. *Bianchi, L.*: Lezioni di geometria differenziale, vol. 2. Zanichelli. Bologna.

7. *Schouten, J. A.*: Über die Bianchische Identität für symmetrische Übertragungen. Math. Zeitschrift, vol. 17, pp. 111–115.

8. *Veblen, O.*: Equiaffine geometry of paths. Proceedings of Nat. Acad. of Sciences, vol. 9, pp. 3, 4.

9. *Eisenhart, L. P.*: Affine geometries of paths possessing an invariant integral. Proceedings of Nat. Acad. of Sciences, vol. 9, pp. 4–7.

1924. 1. *Schouten, J. A.*: Der Ricci-Kalkül. Springer, Berlin.

2. *Eisenhart, L. P.*: Geometries of paths for which the equations of the paths admit a quadratic first integral. Transactions of the Amer. Math. Soc., vol. 26, pp. 378–384.

3. *Cartan, E.*: Sur les variétés a connexion projective. Bulletin de la Soc. Math. de France, vol. 52, pp. 205–241.

1925. 1. *Friesecke, H.*: Vektorübertragung, Richtungsübertragung, Metrik. Math. Annalen, vol. 94, pp. 101–118.

2. *Thomas, T. Y.*: On the projective and equiprojective geometries of paths. Proceedings of Nat. Acad. of Sciences, vol. 11, pp. 199–203.

3. *Thomas, J. M.*: Note on the projective geometry of paths. Proceedings of Nat. Acad. of Sciences, vol. 11, pp. 207–209.

4. *Veblen, O.* and *Thomas, J. M.*: Projective normal coördinates for the geometry of paths. Proceedings of Nat. Acad. of Sciences, vol 11., pp. 204–207.

5. *Levi-Civita, T.*: Lezioni di calcolo differenziale assoluto. Stock, Roma.

6. *Schouten, J. A.*: On the conditions of integrability of covariant differential equations. Transactions of the Amer. Math. Soc., vol. 27, pp. 441–473.

7. *Veblen, O.*: Remarks on the foundations of geometry. Bulletin of the Amer. Math. Soc., vol. 31, pp. 121–141.

8. *Synge, J. L.*: A generalization of the Riemannian line-element. Transactions of the Amer. Math. Soc., vol. 27, pp. 61–67.

9. *Taylor, J. H.*: A generalization of Levi-Civita's parallelism and the Frenet formulas. Transactions of the Amer. Math. Soc., vol. 27, pp. 246–264.

10. *Thomas, T. Y.*: Note on the projective geometry of paths. Bulletin of the Amer. Math. Soc., vol. 31, pp. 318–322.

11. *Einstein, A.*: Einheitliche Feldtheorie von Gravitation und Elektrizität. Sitzungsber. der Preuß. Akad. der Wissensch. zu Berlin, pp. 414-419.

1926. 1. *Eisenhart, L. P.*: Riemannian Geometry. Princeton University Press.

2. *Berwald, L.*: Über Parallelübertragung in Räumen mit allgemeiner Maßbestimmung. Jahresber. Deut. Math. Vereinigung, vol. 24, pp. 213–220.

3. *Thomas, J. M.*: Asymmetric displacement of a vector. Transactions of the Amer. Math. Soc., vol. 28, pp. 658–670.

4. *Levi-Civita, T.*: Sur l'écart géodésique. Math. Annalen, vol. 97, pp. 291–320.

5. *Levy, H.*: Symmetric tensors of the second order whose covariant derivatives vanish. Annals of Mathematics, ser. 2, vol. 27, pp. 91–98.

6. *Veblen, O.* and *Thomas, J. M.*: Projective invariants of affine geometry of paths. Annals of Mathematics, ser. 2, vol. 27, pp. 279–296.

7. *Thomas, J. M.*: First integrals in the geometry of paths. Proceedings of Nat. Acad. of Sciences, vol. 12, pp. 117–124.

8. *Thomas, J. M.*: On normal coördinates in the geometry of paths. Proceedings of Nat. Acad. of Sciences, vol. 12, pp. 58–63.

9. *Eisenhart, L. P.*: Geometries of paths for which the equations of the paths admit $n(n+1)/2$ independent linear first integrals. Transactions of the Amer. Math. Soc., vol. 28, pp. 330–338.

10. *Thomas, T. Y.*: A projective theory of affinely connected manifolds. Math. Zeitschrift, vol. 25, pp. 723–733.

11. *Berwald, L.*: Untersuchung der Krümmung allgemeiner metrischer Räume auf Grund des in ihnen herrschenden Parallelismus. Math. Zeitschrift, vol. 25, pp. 40-73.

12. *Cartan, E.* and *Schouten, J. A.*: On Riemannian geometries admitting an absolute parallelism. Proceedings Kon. Akad. v. Wetenschaffen Amsterdam, vol 29, pp. 933–946.

13. *Thomas, J. M.*: On various geometries giving a unified electric and gravitational theory. Proceedings of Nat. Acad. of Sciences, vol. 12, pp. 187–191.

14. *Eisenhart, L. P.*: Congruences of parallelism of a field of vectors, Proceedings of Nat. Acad. of Sciences, vol. 12, pp. 757-760.

1927. 1. *Levy, H.*: Congruences of curves in the geometry of paths. Rendiconti di Palermo, vol. 51, pp. 304–311.

2. *Eisenhart, L. P.* and *Knebelman, M. S.*: Displacements in a geometry of paths which carry paths into paths. Proceedings of Nat. Acad. of Sciences, vol. 13, pp. 38–42.

3. *Thomas, T. Y.* and *Michal, A. D.*: Differential invariants of relative quadratic differential forms. Annals of Mathematics, ser. 2, vol. 28, pp. 631–688.

4. *Knebelman, M. S.*: Groups of collineations in a space of paths. Proceedings of Nat. Acad. of Sciences, vol. 13, pp. 396–400.

1928. 1. *Douglas, J.*: The general geometry of paths. Annals of Mathematics, ser. 2, vol. 29.

A CATALOG OF SELECTED

DOVER BOOKS

IN SCIENCE AND MATHEMATICS

Astronomy

BURNHAM'S CELESTIAL HANDBOOK, Robert Burnham, Jr. Thorough guide to the stars beyond our solar system. Exhaustive treatment. Alphabetical by constellation: Andromeda to Cetus in Vol. 1; Chamaeleon to Orion in Vol. 2; and Pavo to Vulpecula in Vol. 3. Hundreds of illustrations. Index in Vol. 3. 2,000pp. 6⅛ x 9¼.
Vol. I: 23567-X
Vol. II: 23568-8
Vol. III: 23673-0

EXPLORING THE MOON THROUGH BINOCULARS AND SMALL TELESCOPES, Ernest H. Cherrington, Jr. Informative, profusely illustrated guide to locating and identifying craters, rills, seas, mountains, other lunar features. Newly revised and updated with special section of new photos. Over 100 photos and diagrams. 240pp. 8¼ x 11. 24491-1

THE EXTRATERRESTRIAL LIFE DEBATE, 1750–1900, Michael J. Crowe. First detailed, scholarly study in English of the many ideas that developed from 1750 to 1900 regarding the existence of intelligent extraterrestrial life. Examines ideas of Kant, Herschel, Voltaire, Percival Lowell, many other scientists and thinkers. 16 illustrations. 704pp. 5⅜ x 8½. 40675-X

THEORIES OF THE WORLD FROM ANTIQUITY TO THE COPERNICAN REVOLUTION, Michael J. Crowe. Newly revised edition of an accessible, enlightening book recreates the change from an earth-centered to a sun-centered conception of the solar system. 242pp. 5⅜ x 8½. 41444-2

A HISTORY OF ASTRONOMY, A. Pannekoek. Well-balanced, carefully reasoned study covers such topics as Ptolemaic theory, work of Copernicus, Kepler, Newton, Eddington's work on stars, much more. Illustrated. References. 521pp. 5⅜ x 8½.
65994-1

A COMPLETE MANUAL OF AMATEUR ASTRONOMY: Tools and Techniques for Astronomical Observations, P. Clay Sherrod with Thomas L. Koed. Concise, highly readable book discusses: selecting, setting up and maintaining a telescope; amateur studies of the sun; lunar topography and occultations; observations of Mars, Jupiter, Saturn, the minor planets and the stars; an introduction to photoelectric photometry; more. 1981 ed. 124 figures. 26 halftones. 37 tables. 335pp. 6½ x 9¼.
42820-6

AMATEUR ASTRONOMER'S HANDBOOK, J. B. Sidgwick. Timeless, comprehensive coverage of telescopes, mirrors, lenses, mountings, telescope drives, micrometers, spectroscopes, more. 189 illustrations. 576pp. 5⅝ x 8¼. (Available in U.S. only.)
24034-7

STARS AND RELATIVITY, Ya. B. Zel'dovich and I. D. Novikov. Vol. 1 of *Relativistic Astrophysics* by famed Russian scientists. General relativity, properties of matter under astrophysical conditions, stars, and stellar systems. Deep physical insights, clear presentation. 1971 edition. References. 544pp. 5⅝ x 8¼. 69424-0

Chemistry

THE SCEPTICAL CHYMIST: The Classic 1661 Text, Robert Boyle. Boyle defines the term "element," asserting that all natural phenomena can be explained by the motion and organization of primary particles. 1911 ed. viii+232pp. 5⅜ x 8½. 42825-7

RADIOACTIVE SUBSTANCES, Marie Curie. Here is the celebrated scientist's doctoral thesis, the prelude to her receipt of the 1903 Nobel Prize. Curie discusses establishing atomic character of radioactivity found in compounds of uranium and thorium; extraction from pitchblende of polonium and radium; isolation of pure radium chloride; determination of atomic weight of radium; plus electric, photographic, luminous, heat, color effects of radioactivity. ii+94pp. 5⅜ x 8½. 42550-9

CHEMICAL MAGIC, Leonard A. Ford. Second Edition, Revised by E. Winston Grundmeier. Over 100 unusual stunts demonstrating cold fire, dust explosions, much more. Text explains scientific principles and stresses safety precautions. 128pp. 5⅜ x 8½. 67628-5

THE DEVELOPMENT OF MODERN CHEMISTRY, Aaron J. Ihde. Authoritative history of chemistry from ancient Greek theory to 20th-century innovation. Covers major chemists and their discoveries. 209 illustrations. 14 tables. Bibliographies. Indices. Appendices. 851pp. 5⅜ x 8½. 64235-6

CATALYSIS IN CHEMISTRY AND ENZYMOLOGY, William P. Jencks. Exceptionally clear coverage of mechanisms for catalysis, forces in aqueous solution, carbonyl- and acyl-group reactions, practical kinetics, more. 864pp. 5⅜ x 8½. 65460-5

ELEMENTS OF CHEMISTRY, Antoine Lavoisier. Monumental classic by founder of modern chemistry in remarkable reprint of rare 1790 Kerr translation. A must for every student of chemistry or the history of science. 539pp. 5⅜ x 8½. 64624-6

THE HISTORICAL BACKGROUND OF CHEMISTRY, Henry M. Leicester. Evolution of ideas, not individual biography. Concentrates on formulation of a coherent set of chemical laws. 260pp. 5⅜ x 8½. 61053-5

A SHORT HISTORY OF CHEMISTRY, J. R. Partington. Classic exposition explores origins of chemistry, alchemy, early medical chemistry, nature of atmosphere, theory of valency, laws and structure of atomic theory, much more. 428pp. 5⅜ x 8½. (Available in U.S. only.) 65977-1

GENERAL CHEMISTRY, Linus Pauling. Revised 3rd edition of classic first-year text by Nobel laureate. Atomic and molecular structure, quantum mechanics, statistical mechanics, thermodynamics correlated with descriptive chemistry. Problems. 992pp. 5⅜ x 8½. 65622-5

FROM ALCHEMY TO CHEMISTRY, John Read. Broad, humanistic treatment focuses on great figures of chemistry and ideas that revolutionized the science. 50 illustrations. 240pp. 5⅜ x 8½. 28690-8

Engineering

DE RE METALLICA, Georgius Agricola. The famous Hoover translation of greatest treatise on technological chemistry, engineering, geology, mining of early modern times (1556). All 289 original woodcuts. 638pp. 6¾ x 11. 60006-8

FUNDAMENTALS OF ASTRODYNAMICS, Roger Bate et al. Modern approach developed by U.S. Air Force Academy. Designed as a first course. Problems, exercises. Numerous illustrations. 455pp. 5⅜ x 8½. 60061-0

DYNAMICS OF FLUIDS IN POROUS MEDIA, Jacob Bear. For advanced students of ground water hydrology, soil mechanics and physics, drainage and irrigation engineering, and more. 335 illustrations. Exercises, with answers. 784pp. 6⅛ x 9¼. 65675-6

THEORY OF VISCOELASTICITY (Second Edition), Richard M. Christensen. Complete, consistent description of the linear theory of the viscoelastic behavior of materials. Problem-solving techniques discussed. 1982 edition. 29 figures. xiv+364pp. 6⅛ x 9¼. 42880-X

MECHANICS, J. P. Den Hartog. A classic introductory text or refresher. Hundreds of applications and design problems illuminate fundamentals of trusses, loaded beams and cables, etc. 334 answered problems. 462pp. 5⅜ x 8½. 60754-2

MECHANICAL VIBRATIONS, J. P. Den Hartog. Classic textbook offers lucid explanations and illustrative models, applying theories of vibrations to a variety of practical industrial engineering problems. Numerous figures. 233 problems, solutions. Appendix. Index. Preface. 436pp. 5⅜ x 8½. 64785-4

STRENGTH OF MATERIALS, J. P. Den Hartog. Full, clear treatment of basic material (tension, torsion, bending, etc.) plus advanced material on engineering methods, applications. 350 answered problems. 323pp. 5⅜ x 8½. 60755-0

A HISTORY OF MECHANICS, René Dugas. Monumental study of mechanical principles from antiquity to quantum mechanics. Contributions of ancient Greeks, Galileo, Leonardo, Kepler, Lagrange, many others. 671pp. 5⅜ x 8½. 65632-2

STABILITY THEORY AND ITS APPLICATIONS TO STRUCTURAL MECHANICS, Clive L. Dym. Self-contained text focuses on Koiter postbuckling analyses, with mathematical notions of stability of motion. Basing minimum energy principles for static stability upon dynamic concepts of stability of motion, it develops asymptotic buckling and postbuckling analyses from potential energy considerations, with applications to columns, plates, and arches. 1974 ed. 208pp. 5⅜ x 8½. 42541-X

METAL FATIGUE, N. E. Frost, K. J. Marsh, and L. P. Pook. Definitive, clearly written, and well-illustrated volume addresses all aspects of the subject, from the historical development of understanding metal fatigue to vital concepts of the cyclic stress that causes a crack to grow. Includes 7 appendixes. 544pp. 5⅜ x 8½. 40927-9

Mathematics

FUNCTIONAL ANALYSIS (Second Corrected Edition), George Bachman and Lawrence Narici. Excellent treatment of subject geared toward students with background in linear algebra, advanced calculus, physics, and engineering. Text covers introduction to inner-product spaces, normed, metric spaces, and topological spaces; complete orthonormal sets, the Hahn-Banach Theorem and its consequences, and many other related subjects. 1966 ed. 544pp. 6⅛ x 9¼. 40251-7

ASYMPTOTIC EXPANSIONS OF INTEGRALS, Norman Bleistein & Richard A. Handelsman. Best introduction to important field with applications in a variety of scientific disciplines. New preface. Problems. Diagrams. Tables. Bibliography. Index. 448pp. 5⅜ x 8½. 65082-0

VECTOR AND TENSOR ANALYSIS WITH APPLICATIONS, A. I. Borisenko and I. E. Tarapov. Concise introduction. Worked-out problems, solutions, exercises. 257pp. 5⅝ x 8¼. 63833-2

THE ABSOLUTE DIFFERENTIAL CALCULUS (CALCULUS OF TENSORS), Tullio Levi-Civita. Great 20th-century mathematician's classic work on material necessary for mathematical grasp of theory of relativity. 452pp. 5⅝ x 8¼. 63401-9

AN INTRODUCTION TO ORDINARY DIFFERENTIAL EQUATIONS, Earl A. Coddington. A thorough and systematic first course in elementary differential equations for undergraduates in mathematics and science, with many exercises and problems (with answers). Index. 304pp. 5⅜ x 8½. 65942-9

FOURIER SERIES AND ORTHOGONAL FUNCTIONS, Harry F. Davis. An incisive text combining theory and practical example to introduce Fourier series, orthogonal functions and applications of the Fourier method to boundary-value problems. 570 exercises. Answers and notes. 416pp. 5⅜ x 8½. 65973-9

COMPUTABILITY AND UNSOLVABILITY, Martin Davis. Classic graduate-level introduction to theory of computability, usually referred to as theory of recurrent functions. New preface and appendix. 288pp. 5⅜ x 8½. 61471-9

ASYMPTOTIC METHODS IN ANALYSIS, N. G. de Bruijn. An inexpensive, comprehensive guide to asymptotic methods–the pioneering work that teaches by explaining worked examples in detail. Index. 224pp. 5⅜ x 8½ 64221-6

APPLIED COMPLEX VARIABLES, John W. Dettman. Step-by-step coverage of fundamentals of analytic function theory–plus lucid exposition of five important applications: Potential Theory; Ordinary Differential Equations; Fourier Transforms; Laplace Transforms; Asymptotic Expansions. 66 figures. Exercises at chapter ends. 512pp. 5⅜ x 8½. 64670-X

INTRODUCTION TO LINEAR ALGEBRA AND DIFFERENTIAL EQUATIONS, John W. Dettman. Excellent text covers complex numbers, determinants, orthonormal bases, Laplace transforms, much more. Exercises with solutions. Undergraduate level. 416pp. 5⅜ x 8½. 65191-6

CALCULUS OF VARIATIONS WITH APPLICATIONS, George M. Ewing. Applications-oriented introduction to variational theory develops insight and promotes understanding of specialized books, research papers. Suitable for advanced undergraduate/graduate students as primary, supplementary text. 352pp. 5⅜ x 8½. 64856-7

COMPLEX VARIABLES, Francis J. Flanigan. Unusual approach, delaying complex algebra till harmonic functions have been analyzed from real variable viewpoint. Includes problems with answers. 364pp. 5⅜ x 8½. 61388-7

AN INTRODUCTION TO THE CALCULUS OF VARIATIONS, Charles Fox. Graduate-level text covers variations of an integral, isoperimetrical problems, least action, special relativity, approximations, more. References. 279pp. 5⅜ x 8½. 65499-0

COUNTEREXAMPLES IN ANALYSIS, Bernard R. Gelbaum and John M. H. Olmsted. These counterexamples deal mostly with the part of analysis known as "real variables." The first half covers the real number system, and the second half encompasses higher dimensions. 1962 edition. xxiv+198pp. 5⅜ x 8½. 42875-3

CATASTROPHE THEORY FOR SCIENTISTS AND ENGINEERS, Robert Gilmore. Advanced-level treatment describes mathematics of theory grounded in the work of Poincaré, R. Thom, other mathematicians. Also important applications to problems in mathematics, physics, chemistry, and engineering. 1981 edition. References. 28 tables. 397 black-and-white illustrations. xvii+666pp. 6⅛ x 9¼. 67539-4

INTRODUCTION TO DIFFERENCE EQUATIONS, Samuel Goldberg. Exceptionally clear exposition of important discipline with applications to sociology, psychology, economics. Many illustrative examples; over 250 problems. 260pp. 5⅜ x 8½. 65084-7

NUMERICAL METHODS FOR SCIENTISTS AND ENGINEERS, Richard Hamming. Classic text stresses frequency approach in coverage of algorithms, polynomial approximation, Fourier approximation, exponential approximation, other topics. Revised and enlarged 2nd edition. 721pp. 5⅜ x 8½. 65241-6

INTRODUCTION TO NUMERICAL ANALYSIS (2nd Edition), F. B. Hildebrand. Classic, fundamental treatment covers computation, approximation, interpolation, numerical differentiation and integration, other topics. 150 new problems. 669pp. 5⅜ x 8½. 65363-3

THREE PEARLS OF NUMBER THEORY, A. Y. Khinchin. Three compelling puzzles require proof of a basic law governing the world of numbers. Challenges concern van der Waerden's theorem, the Landau-Schnirelmann hypothesis and Mann's theorem, and a solution to Waring's problem. Solutions included. 64pp. 5⅜ x 8½. 40026-3

THE PHILOSOPHY OF MATHEMATICS: An Introductory Essay, Stephan Körner. Surveys the views of Plato, Aristotle, Leibniz & Kant concerning propositions and theories of applied and pure mathematics. Introduction. Two appendices. Index. 198pp. 5⅜ x 8½. 25048-2

Math–Decision Theory, Statistics, Probability

Math–Geometry and Topology

ELEMENTARY CONCEPTS OF TOPOLOGY, Paul Alexandroff. Elegant, intuitive approach to topology from set-theoretic topology to Betti groups; how concepts of topology are useful in math and physics. 25 figures. 57pp. 5⅜ x 8½. 60747-X

COMBINATORIAL TOPOLOGY, P. S. Alexandrov. Clearly written, well-organized, three-part text begins by dealing with certain classic problems without using the formal techniques of homology theory and advances to the central concept, the Betti groups. Numerous detailed examples. 654pp. 5⅜ x 8½. 40179-0

EXPERIMENTS IN TOPOLOGY, Stephen Barr. Classic, lively explanation of one of the byways of mathematics. Klein bottles, Moebius strips, projective planes, map coloring, problem of the Koenigsberg bridges, much more, described with clarity and wit. 43 figures. 210pp. 5⅜ x 8½. 25933-1

CONFORMAL MAPPING ON RIEMANN SURFACES, Harvey Cohn. Lucid, insightful book presents ideal coverage of subject. 334 exercises make book perfect for self-study. 55 figures. 352pp. 5⅝ x 8¼. 64025-6

THE GEOMETRY OF RENÉ DESCARTES, René Descartes. The great work founded analytical geometry. Original French text, Descartes's own diagrams, together with definitive Smith-Latham translation. 244pp. 5⅜ x 8½. 60068-8

PRACTICAL CONIC SECTIONS: The Geometric Properties of Ellipses, Parabolas and Hyperbolas, J. W. Downs. This text shows how to create ellipses, parabolas, and hyperbolas. It also presents historical background on their ancient origins and describes the reflective properties and roles of curves in design applications. 1993 ed. 98 figures. xii+100pp. 6½ x 9¼. 42876-1

THE THIRTEEN BOOKS OF EUCLID'S ELEMENTS, translated with introduction and commentary by Thomas L. Heath. Definitive edition. Textual and linguistic notes, mathematical analysis. 2,500 years of critical commentary. Unabridged. 1,414pp. 5⅜ x 8½.. Three-vol. set. Vol. I: 60088-2 Vol. II: 60089-0 Vol. III: 60090-4

GEOMETRY OF COMPLEX NUMBERS, Hans Schwerdtfeger. Illuminating, widely praised book on analytic geometry of circles, the Moebius transformation, and two-dimensional non-Euclidean geometries. 200pp. 5⅝ x 8¼. 63830-8

DIFFERENTIAL GEOMETRY, Heinrich W. Guggenheimer. Local differential geometry as an application of advanced calculus and linear algebra. Curvature, transformation groups, surfaces, more. Exercises. 62 figures. 378pp. 5⅜ x 8½. 63433-7

CURVATURE AND HOMOLOGY: Enlarged Edition, Samuel I. Goldberg. Revised edition examines topology of differentiable manifolds; curvature, homology of Riemannian manifolds; compact Lie groups; complex manifolds; curvature, homology of Kaehler manifolds. New Preface. Four new appendixes. 416pp. 5⅜ x 8½. 40207-X

History of Math

THE WORKS OF ARCHIMEDES, Archimedes (T. L. Heath, ed.). Topics include the famous problems of the ratio of the areas of a cylinder and an inscribed sphere; the measurement of a circle; the properties of conoids, spheroids, and spirals; and the quadrature of the parabola. Informative introduction. clxxxvi+326pp; supplement, 52pp. 5⅜ x 8½. 42084-1

A SHORT ACCOUNT OF THE HISTORY OF MATHEMATICS, W. W. Rouse Ball. One of clearest, most authoritative surveys from the Egyptians and Phoenicians through 19th-century figures such as Grassman, Galois, Riemann. Fourth edition. 522pp. 5⅜ x 8½. 20630-0

THE HISTORY OF THE CALCULUS AND ITS CONCEPTUAL DEVELOPMENT, Carl B. Boyer. Origins in antiquity, medieval contributions, work of Newton, Leibniz, rigorous formulation. Treatment is verbal. 346pp. 5⅜ x 8½. 60509-4

THE HISTORICAL ROOTS OF ELEMENTARY MATHEMATICS, Lucas N. H. Bunt, Phillip S. Jones, and Jack D. Bedient. Fundamental underpinnings of modern arithmetic, algebra, geometry, and number systems derived from ancient civilizations. 320pp. 5⅜ x 8½. 25563-8

A HISTORY OF MATHEMATICAL NOTATIONS, Florian Cajori. This classic study notes the first appearance of a mathematical symbol and its origin, the competition it encountered, its spread among writers in different countries, its rise to popularity, its eventual decline or ultimate survival. Original 1929 two-volume edition presented here in one volume. xxviii+820pp. 5⅜ x 8½. 67766-4

GAMES, GODS & GAMBLING: A History of Probability and Statistical Ideas, F. N. David. Episodes from the lives of Galileo, Fermat, Pascal, and others illustrate this fascinating account of the roots of mathematics. Features thought-provoking references to classics, archaeology, biography, poetry. 1962 edition. 304pp. 5⅜ x 8½. (Available in U.S. only.) 40023-9

OF MEN AND NUMBERS: The Story of the Great Mathematicians, Jane Muir. Fascinating accounts of the lives and accomplishments of history's greatest mathematical minds–Pythagoras, Descartes, Euler, Pascal, Cantor, many more. Anecdotal, illuminating. 30 diagrams. Bibliography. 256pp. 5⅜ x 8½. 28973-7

HISTORY OF MATHEMATICS, David E. Smith. Nontechnical survey from ancient Greece and Orient to late 19th century; evolution of arithmetic, geometry, trigonometry, calculating devices, algebra, the calculus. 362 illustrations. 1,355pp. 5⅜ x 8½. Two-vol. set. Vol. I: 20429-4 Vol. II: 20430-8

A CONCISE HISTORY OF MATHEMATICS, Dirk J. Struik. The best brief history of mathematics. Stresses origins and covers every major figure from ancient Near East to 19th century. 41 illustrations. 195pp. 5⅜ x 8½. 60255-9

Physics

OPTICAL RESONANCE AND TWO-LEVEL ATOMS, L. Allen and J. H. Eberly. Clear, comprehensive introduction to basic principles behind all quantum optical resonance phenomena. 53 illustrations. Preface. Index. 256pp. 5⅜ x 8½. 65533-4

QUANTUM THEORY, David Bohm. This advanced undergraduate-level text presents the quantum theory in terms of qualitative and imaginative concepts, followed by specific applications worked out in mathematical detail. Preface. Index. 655pp. 5⅜ x 8½. 65969-0

ATOMIC PHYSICS: 8th edition, Max Born. Nobel laureate's lucid treatment of kinetic theory of gases, elementary particles, nuclear atom, wave-corpuscles, atomic structure and spectral lines, much more. Over 40 appendices, bibliography. 495pp. 5⅜ x 8½. 65984-4

A SOPHISTICATE'S PRIMER OF RELATIVITY, P. W. Bridgman. Geared toward readers already acquainted with special relativity, this book transcends the view of theory as a working tool to answer natural questions: What is a frame of reference? What is a "law of nature"? What is the role of the "observer"? Extensive treatment, written in terms accessible to those without a scientific background. 1983 ed. xlviii+172pp. 5⅜ x 8½. 42549-5

AN INTRODUCTION TO HAMILTONIAN OPTICS, H. A. Buchdahl. Detailed account of the Hamiltonian treatment of aberration theory in geometrical optics. Many classes of optical systems defined in terms of the symmetries they possess. Problems with detailed solutions. 1970 edition. xv+360pp. 5⅜ x 8½. 67597-1

PRIMER OF QUANTUM MECHANICS, Marvin Chester. Introductory text examines the classical quantum bead on a track: its state and representations; operator eigenvalues; harmonic oscillator and bound bead in a symmetric force field; and bead in a spherical shell. Other topics include spin, matrices, and the structure of quantum mechanics; the simplest atom; indistinguishable particles; and stationary-state perturbation theory. 1992 ed. xiv+314pp. 6⅛ x 9¼. 42878-8

LECTURES ON QUANTUM MECHANICS, Paul A. M. Dirac. Four concise, brilliant lectures on mathematical methods in quantum mechanics from Nobel Prize–winning quantum pioneer build on idea of visualizing quantum theory through the use of classical mechanics. 96pp. 5⅜ x 8½. 41713-1

THIRTY YEARS THAT SHOOK PHYSICS: The Story of Quantum Theory, George Gamow. Lucid, accessible introduction to influential theory of energy and matter. Careful explanations of Dirac's anti-particles, Bohr's model of the atom, much more. 12 plates. Numerous drawings. 240pp. 5⅜ x 8½. 24895-X

ELECTRONIC STRUCTURE AND THE PROPERTIES OF SOLIDS: The Physics of the Chemical Bond, Walter A. Harrison. Innovative text offers basic understanding of the electronic structure of covalent and ionic solids, simple metals, transition metals and their compounds. Problems. 1980 edition. 582pp. 6⅛ x 9¼. 66021-4

QUANTUM MECHANICS: Principles and Formalism, Roy McWeeny. Graduate student–oriented volume develops subject as fundamental discipline, opening with review of origins of Schrödinger's equations and vector spaces. Focusing on main principles of quantum mechanics and their immediate consequences, it concludes with final generalizations covering alternative "languages" or representations. 1972 ed. 15 figures. xi+155pp. 5⅜ x 8½. 42829-X

INTRODUCTION TO QUANTUM MECHANICS WITH APPLICATIONS TO CHEMISTRY, Linus Pauling & E. Bright Wilson, Jr. Classic undergraduate text by Nobel Prize winner applies quantum mechanics to chemical and physical problems. Numerous tables and figures enhance the text. Chapter bibliographies. Appendices. Index. 468pp. 5⅜ x 8½. 64871-0

METHODS OF THERMODYNAMICS, Howard Reiss. Outstanding text focuses on physical technique of thermodynamics, typical problem areas of understanding, and significance and use of thermodynamic potential. 1965 edition. 238pp. 5⅜ x 8½. 69445-3

TENSOR ANALYSIS FOR PHYSICISTS, J. A. Schouten. Concise exposition of the mathematical basis of tensor analysis, integrated with well-chosen physical examples of the theory. Exercises. Index. Bibliography. 289pp. 5⅜ x 8½. 65582-2

THE ELECTROMAGNETIC FIELD, Albert Shadowitz. Comprehensive undergraduate text covers basics of electric and magnetic fields, builds up to electromagnetic theory. Also related topics, including relativity. Over 900 problems. 768pp. 5⅝ x 8¼. 65660-8

GREAT EXPERIMENTS IN PHYSICS: Firsthand Accounts from Galileo to Einstein, Morris H. Shamos (ed.). 25 crucial discoveries: Newton's laws of motion, Chadwick's study of the neutron, Hertz on electromagnetic waves, more. Original accounts clearly annotated. 370pp. 5⅜ x 8½. 25346-5

RELATIVITY, THERMODYNAMICS AND COSMOLOGY, Richard C. Tolman. Landmark study extends thermodynamics to special, general relativity; also applications of relativistic mechanics, thermodynamics to cosmological models. 501pp. 5⅜ x 8½. 65383-8

STATISTICAL PHYSICS, Gregory H. Wannier. Classic text combines thermodynamics, statistical mechanics, and kinetic theory in one unified presentation of thermal physics. Problems with solutions. Bibliography. 532pp. 5⅜ x 8½. 65401-X